RICHARD DAWKINS

HOW A SCIENTIST CHANGED
THE WAY WE THINK

RICHARD DAWKINS

HOW A SCIENTIST CHANGED THE WAY WE THINK

Reflections by scientists, writers, and philosophers

Edited by

ALAN GRAFEN

AND

MARK RIDLEY

OXFORD

UNIVERSITY PRESS

OXFORD

UNIVERSITY PRESS

Great Clarendon Street, Oxford OX2 6DP

Oxford University Press is a department of the University of Oxford.
It furthers the University's objective of excellence in research, scholarship,
and education by publishing worldwide in

Oxford New York

Auckland Cape Town Dar es Salaam Hong Kong Karachi
Kuala Lumpur Madrid Melbourne Mexico City Nairobi
New Delhi Shanghai Taipei Toronto

With offices in

Argentina Austria Brazil Chile Czech Republic France Greece
Guatemala Hungary Italy Japan Poland Portugal Singapore
South Korea Switzerland Thailand Turkey Ukraine Vietnam

Oxford is a registered trade mark of Oxford University Press
in the UK and in certain other countries

Published in the United States
by Oxford University Press Inc., New York

British Library Cataloguing in Publication Data

Data available

Library of Congress Cataloging in Publication Data

Data available

Typeset by RefineCatch Ltd., Bungay, Suffolk
Printed in Great Britain by
Clays Ltd., St Ives plc

ISBN 0-19-929116-0 978-0-19-929116-8

1

CONTENTS

LOGIC

ANTIPHONAL VOICES

HUMANS

CONTROVERSY

WRITING

LIST OF CONTRIBUTORS

Robert Aunger, Senior Lecturer in Evolutionary Public Health, London School of Hygiene and Tropical Medicine, author of *The Electric Meme*, and editor of *Darwinizing Culture*.

David P. Barash, Professor of Psychology, University of Washington, author of *Madame Bovary's Ovaries: A Darwinian look at literature*, *The Myth of Monogamy*, *Revolutionary Biology*, and other books.

Sir Patrick Bateson, FRS. Emeritus Professor of Ethology, Cambridge University, co-author of *Design for a Life: How Behaviour Develops*.

Seth Bullock, Senior Lecturer in the Science and Engineering of Natural Systems, School of Electronics and Computer Science, University of Southampton.

Helena Cronin, Co-Director, London School of Economics Centre for Philosophy of Natural and Social Science, organizer of the Darwin@LSE programme, and author of *The Ant and the Peacock*.

Martin Daly, Professor of Psychology, McMaster University, Canada, co-author of *The Truth About Cinderella* and other books.

Marian Stamp Dawkins, Fellow in Biological Sciences at Somerville College and Professor of Animal Behaviour, Zoology Department, Oxford University, author of *Through Our Eyes Only* and *Animal Suffering*.

Daniel C. Dennett, Professor of Philosophy, Tufts University, author of *Darwin's Dangerous Idea*, *Consciousness Explained*, *Freedom Evolves*, and other books.

David Deutsch, Professor, Centre for Quantum Computing, Clarendon Laboratory, Oxford University, author of *The Fabric of Reality*.

Alan Grafen, Professor of Theoretical Biology, Oxford University.

A. C. Grayling, Professor, Department of Philosophy, Birkbeck College, London University, author of *The Meaning of Things*, *What is Good?*, *Life, Sex and Ideas: The Good Life Without God*, *Philosophy*, and other books.

David Haig, Professor, Department of Organismic and Evolutionary Biology, Harvard University, author of *Genomic Imprinting and Kinship*.

Michael Hansell, Professor, Institute of Biomedical and Life Sciences, Glasgow University, author of *Animal Architecture* and other books.

The Rt Revd Richard Harries, Bishop of Oxford, author of *Art and the Beauty of God*, *God Outside the Box*, and other books.

Sir John Krebs, FRS. Professor, Department of Zoology, and Principal, Jesus College, Oxford University; former chairman, Food Standards Agency.

Marek Kohn, Visiting Fellow at the School of Life Sciences, University of Sussex, author of *A Reason for Everything*, *As We Know It: Coming to Terms with an Evolved Mind*, *The Race Gallery*, and other books.

Randolph M. Nesse, Professor of Psychiatry and Professor of Psychology, University of Michigan, author of *Why We Get Sick* and other books.

Steven Pinker, Johnstone Family Professor, Department of Psychology, Harvard University, author of *The Language Instinct*, *How the Mind Works*, *The Blank Slate*, and other books.

Philip Pullman, author of *His Dark Materials* trilogy, *Ruby in the Smoke*, and other books.

Andrew F. Read, Professor of Natural History, Edinburgh University.

Matt Ridley, writer and former journalist, author of *Nature via Nurture*, *Genome*, *The Red Queen*, and of a forthcoming biography of Francis Crick.

Michael Ruse, Professor of Philosophy, Florida State University, author of *The Evolution–Creation Struggle*, *Darwin and Design*, and other books.

Ullica Segerstråle, Professor of Sociology, Illinois Institute of Technology, Chicago, author of *Defenders of the Truth*, and editor of other books.

Michael Shermer, founding publisher of *Skeptic* magazine, contributing editor of *Scientific American*, author of *How We Believe: The Science of Good and Evil*, and *In Darwin's Shadow: The Life of Alfred Russel Wallace*.

Kim Sterelny, Professor of Philosophy, Victoria University, Wellington, and the ANU, Canberra, author of *Thought in a Hostile World: The Evolution of Human Cognition*, *Dawkins vs Gould: Survival of the Fittest*, and other books.

Margo Wilson, Professor of Psychology, McMaster University, Canada, co-author of *The Truth About Cinderella* and other books.

PREFACE

In 1976, a young Oxford biologist published a book called *The Selfish Gene*. To Richard Dawkins' own surprise and sometimes alarm, it became widely discussed, often misunderstood, and highly influential. *The Selfish Gene* is now well established as a classic exposition of evolutionary ideas for academic and lay readers alike. Its author, propelled to fame, went on to display the range and depth of his analytical skills and literary abilities in a string of best-sellers: *The Extended Phenotype* (intended primarily for fellow biologists), *The Blind Watchmaker*, *River Out of Eden*, *Climbing Mount Improbable*, *Unweaving the Rainbow*, and *The Ancestor's Tale*. A collection of his essays was published as *A Devil's Chaplain*. Increasingly involved in public debate on science and rationalism, Dawkins has become a familiar figure in the media, and a leading champion of atheism. To find professional scientists with a similar public profile on non-scientific issues, one has to return to the days of J. B. S. Haldane and before him T. H. Huxley in the UK, and perhaps Einstein in the USA. In 1995, Dr Charles Simonyi endowed a chair for the Public Understanding of Science at Oxford University that enabled Dawkins, as the first holder, to concentrate on his writing.

This collection of essays considers the range of Dawkins' influence as scientist, writer, and public figure. Inevitably, though, his seminal work, *The Selfish Gene*, takes pride of place among his achievements and thus forms the primary focus of this volume. *The Selfish Gene* reached so many audiences that one person is unlikely to know of them all. The essays of the first few sections illustrate the range of the book's influence, with distinguished authors from many fields explaining how its ideas have affected them personally and professionally.

We begin with accounts by four biologists, whose work in parasitology, gender differences, communication, and animal artefacts has been inspired by Dawkins. The second section focuses on the central text itself and looks at its unfading pedagogical power, its intellectual contribution, and its historical place in the sociobiology debates. In 'The Gene Meme', the Harvard geneticist David Haig contributes a mind-twisting tribute to *The Selfish Gene* by first using the history of the word 'gene' to illustrate the concept of the meme, and then employing that discussion to consider how well the concept of meme stands up.

The most impressive aspect of *The Selfish Gene* is that it is argued from first principles and with complete logic. Logic is the common currency of academe, and it is no surprise that a wide range of thinkers have embraced the book's arguments and gone on to extend them in their own directions. In the third section a philosopher, a computer scientist, a physicist, and a cognitive scientist describe how they have pursued the logic of Dawkins' work, and *The Selfish Gene* in particular, in their respective fields.

Dawkins' writing and public appearances have, of course, engendered considerable controversy and resistance, though much of the latter, as concerns selfish gene ideas, has been rooted in misunderstanding. A book published on the thirtieth anniversary of *The Selfish Gene*, edited by two of his former graduate students, naturally has mainly supporters and defenders as contributors, but we also include some 'antiphonal voices', specifically in the middle section, in which three, admittedly friendly, critics discuss where in their view Dawkins has gone wrong.

The most controversial applications of selfish gene ideas have been to humans, and these form the theme of the fifth section. The issues discussed will be familiar to many readers through their regular irruptions into the popular media.

Dawkins' public profile as a controversialist is the aspect considered in the sixth section, with pieces looking at his role as sceptic and atheist champion, and how his understanding of evolution has influenced his ethics and politics and his views on the meaning of life.

Most of this volume is concerned with the content of Dawkins' books. But the final section considers another fundamental aspect that has made his books such classics—the writing itself.

Dawkins' role in creating a new genre of science writing is assessed. And to conclude the collection, Philip Pullman, who has himself inspired millions of readers through his works of fiction, pays an eloquent tribute to a fellow writer.

As editors, we hope this volume will give valuable insights into the range of work in disparate fields that has been inspired by Dawkins' writings, and demonstrate the extent of his influence. Several of the essays are themselves significant contributions to the scientific debate and should provide pause for thought for professional biologist and general reader alike. We can think of no more fitting tribute to a figure so exactingly logical in science, so patiently lucid in promoting the public understanding of science, and so outspoken and clear-headed in the public sphere.

Alan Grafen
Mark Ridley

BIOLOGY

BALLOONING PARROTS AND SEMI-LUNAR GERMS

Andrew F. Read

P ILLOW talk first introduced me to *The Selfish Gene*. I remember the scene quite vividly, perhaps because of the weather. Sun was streaming into the room and it was exceptionally warm; in New Zealand's southernmost university town, such mornings were rare. I was a second year zoology undergraduate and my then girlfriend was majoring in English literature. I was a great deal more interested in her than in literature, but on this particular morning she told me about a weird biology book she'd had as a set text. I was surprised that a biology book would appear in a literature course, but she said that it was used to discuss the role of metaphor and then said, I believe without irony, that the author proposed that genes had emotions. We both laughed at this lunacy, and I suggested that she should read a sensible evolutionary thinker like Stephen Jay Gould.

Incredible as it now seems, my first physical encounter with the book was *after* I had finished my four year Zoology degree specializing in evolution, ecology, and behaviour. The limitations of my formal education were, I like to think, more than offset by the summer jobs I had with the New Zealand Wildlife Service on remote mountains and offshore islands. The best job came immediately after finals, when I had the extraordinary good fortune to work on the kakapo conservation program during a breeding season. Kakapo are the world's strangest and most fabulous birds. But to me then, as a budding evolutionary biologist, I came to see them as an intellectual affront. I just could not figure out how kakapo could be like they are; I couldn't even figure out how one might figure it out. Certainly, my extensive Stephen Jay Gould collection was no help. My boss recommended *The Selfish Gene*, and this time, I actually read it. What a revelation. It didn't mention kakapo (of course), but here at least was a framework to explain them. I now spend my professional life thinking about infectious diseases the way *The Selfish Gene* taught me to think about kakapo. It turns out that selfish genery not only explains weirdos like kakapo, but also makes insights possible that escape conventional biomedicine.

To explain why, I first need to describe kakapo. They are parrots which break all parrot records. They are the largest (big males weigh the same as a large cat) and the longest lived (an elderly bird found in 1975 is still alive). Beak and feet aside, they do not actually look like parrots: they have an owl-ish face framed with whisker-like feathers. They also do not behave like parrots: they are nocturnal, flightless, and meet only for sex, and then only every three to four years. And sadly, they are one of the world's most endangered parrots. At the time of writing, there are just eighty-six left, the majority of which are males.

Evolution has dealt kakapo an extraordinarily bad hand for life in the modern world. Until the last millennium, the only mammals in New Zealand were bats. Now, of course, there are humans, rats, cats, stoats, and dogs. Like so many of the endemic birds, kakapo were ill-prepared for mammalian hunters. When disturbed, kakapo freeze and hope to blend into the forest background. With their rich green coloration, this is visually very effective. But kakapo smell. They smell so strongly (of sweet hay), even I could sometimes smell them before I could see them. When the freeze routine fails with olfactory hunters, surviving kakapo can only run, climb, or jump away.

Their breeding system is just as hopeless. For most of their lives, kakapo are solitary, spending their time wandering about, grazing, and sleeping. In those rare years when they do breed, males aggregate in breeding arenas which are usually on high promontories. Here, they call out for females living in the surrounding valleys. Even the calling is weird. Males suck air into enlarged air sacs, blowing themselves up so they look like a balloon with a beak. They then deflate themselves, the slow expulsion of air making a low 'sonic' boom. This inflation–deflation cycle, which takes a few minutes, is repeated from the same spot, all night, night after night for up to six months. The eerie boom can be heard for miles, attracting females—and predators. A single cat loose in one of these breeding arenas can wreak havoc.

Females come to these arenas ('leks') to mate, and then leave immediately to resume their solitary lives. Some miles from the leks, they lay eggs in nests built on the forest floor. Each night they go off on foraging trips of up to several kilometres. If the unprotected eggs manage to escape ground-hunting predators for a month, they hatch, only to produce completely helpless chicks. And now, because she has mouths to feed, mum has to spend even more time away foraging. After three months of this, the chicks are finally developed enough to leave the heavily scented nest for the relative safety of the open forest.

It is hard to imagine a lifestyle less suited to withstanding mammalian predators.[1] When I worked with kakapo in the 1980s, all we seemed to do was watch them die. Now they have been shuttled to predator-free islands where their future is brighter, but they will be heavily dependent on human management for some decades, if not in perpetuity.

I was part of a small research team in the south of Stewart Island, the southernmost of the large New Zealand islands, tracking females which had been fitted with radio devices. Alone in a tent on some exposed hilltop, often in the most appalling weather, each of us would take bearings on the handful of birds we were tracking every half hour, all night. Every two weeks the helicopter arrived bearing fresh meat and veg and, occasionally, rapidly melting ice cream and warming beer.

Mud was a feature of the place. The tracks we used to get to our tents were often knee-deep in the stuff. Lying in the tents was like being on a giant dirty waterbed. While we knew someone else was on another hilltop, radio contact was unreliable, and it was difficult not to go mad. Between half-hour telemetry readings we would read or doze (resetting the alarm every half hour). We could only take it for a few nights before needing a break; we used to clamour for the job of sitting in hides at the leks, watching the balloon routine through the night-viewing scopes and, just like the male kakapo, hoping desperately that a female would arrive (in a region where predators had taken all the females, males copulated with fallen tree trunks and even a rolled up sweater). I never saw any females on a lek, but two eggs hatched that year. The chicks didn't last a month. In all of this, I kept asking, why? It is pretty easy to imagine that in a mammal-free world, flight-lessness could evolve: flight is expensive so do without it if you can. The nocturnality (and associated owl-like features), green coloration, and freeze-responses we assumed to be essential adaptations to avoid visual hunters; large eagles existed in New Zealand until quite recently. But this crazy breeding system? Night after night I wondered how such a bizarre behaviour could have evolved. What a stupid thing for a species to do. Surely the males should help feed the chicks? What were they doing mucking around on leks?

In response to my persistent questioning, the scientist in charge, Ralph Powlesland, sent me a paper about other lek breeders which he thought might help. It didn't; it was full of inscrutable mathematics. But maths about breeding systems implied some kind of a theory. I demanded an interpreter and, much to my surprise, *The Selfish Gene* arrived on the next helicopter run.

Lying in my tented, muddy waterbed, I read it by torchlight between half hourly telemetry readings. In fact I read it three times, very slowly, trying to make sense of it all. Reading now the comments I then wrote in the margins, I must have been deeply sceptical (I guess in those days I valued the scientific opinions of English literature students). But there it was, clearly laid out. The good of the species was irrelevant! It was competition between strategies to maximize genetic representation that mattered. This competition could result in outcomes that were disastrous for all. Just as evolution could not have the foresight to arrange kakapo to be prepared for mammals, it could not arrange them to maximize the reproductive output of the species. Individual selfish genes were maximizing their share of the gene pool, even if this meant fewer kakapo offspring overall. The kakapo males must have given up parental care because helping their chicks to survive was not the way for them to maximize their fitness. It didn't matter that the offspring of one mate mostly died: getting more mates must make up for it. And females mating on leks got to choose the best male genes going.

Of course, there in the tent, all this was hypothetical and to be fair, today we still do not fully understand the so-called 'lek paradox' for any species, let alone kakapo. But I think we all agree that the answer is somewhere among the ideas that flow from selfish genery. And what was clear to me then was that here was a framework which had tremendous explanatory power for all of biology. Evolutionary biology could actually *explain* organic diversity—really explain it in a predictive sense, not just describe it. Gould was wrong. Adaptationism could be rigorous, and generate testable ideas, some of which were clearly right. Gone in a stroke the intellectually barren 'it-just-is' hypothesis and woolly group selectionism.

I also drew another conclusion from the book. Much to my surprise, there were clearly people making a living ruminating about stuff I thought about in idle moments. I had already decided to be a career zoologist, but this made me think that perhaps I could—and maybe even should—do something other than applied conservation biology.

Sometime that autumn, a radio message arrived saying that I had won a Ph.D. scholarship to Oxford. Over the next few years in Oxford, I learnt that many people worldwide were involved in working out the logical consequences of *The Selfish Gene*, and that in fact many had been doing so before the book had come out. Indeed, it turned out that the intellectual framework had already been in the air, but *The Selfish Gene* crystallized it and made it impossible to ignore. I learnt that most of the criticisms it attracted were intellectually boring or, worse, stupid. Other frameworks that people proposed as alternatives were either simple rebrandings or vacuous. This was the only show in town, and it was a productive and exciting one. And I also learnt that the impact of the book on me was not unique. Many other students were doing what they were doing because they had chanced upon *The Selfish Gene*. It actually did deserve to be a set text in English Literature 101.

My Ph.D. was mostly concerned with how infectious diseases might be responsible for the bizarre songs, colours, and plumes of many male birds. While I was finishing my thesis, it occurred to me that we evolutionary biologists were fixated on hosts and inexplicably ignoring the infectious disease agents themselves. We had left them to microbiologists and parasitologists, who quite plainly did not think in selfish gene terms. Yet infectious diseases evolve on experimentally measurable timescales, so we could test theory, and because they make us sick, there must be money in it. Various epidemiologists had made forays into disease evolution (most notably Roy Anderson and Bob May, who had looked at the evolution of virulence),[2] and Paul Ewald had been using selfish genery to make controversial claims about the evolution of a swathe of human diseases.[3] But to me, this was but a drop in the ocean of possibilities, and none of it involved the sort of high-class experimental work that flowed from selfish genery, and which was by then captured in John Krebs' and Nick Davies' classic edited volumes and introductory text.[4]

Oxford was, and still is, capable of generating breathtaking intellectual arrogance. This must explain the belief that I developed towards the end of the 1980s that even though I knew absolutely nothing about how to do an experiment, or about any infectious disease, I could demonstrate that not only could selfish genery make sense of facts that biomedicine could not, it could even make novel quantitative predictions that would turn out to be true.

The most heavily annotated part of my original copy of *The Selfish Gene* is Chapter 9, 'Battle of the Sexes'. It was this chapter that really convinced me that the adaptationist programme works. In it, Dawkins lucidly summarizes Fisher's gene-centred explanation of why 1:1 sex ratios are so common, even though they clearly do not maximize the reproductive output of a species (just enough males to fertilize all the females would do this). Fisher's idea was that only a 1:1 sex ratio is evolutionary stable; all others can be invaded by a mutant producing slightly more of the rarer sex. This idea is so logically beautiful that no one bothered to test it experimentally until the 1990s: it simply *had* to be true (it was). Before then, the best evidence that it was true came from species without 1:1 sex ratios. In a 1967 paper, Bill Hamilton showed that where individuals are mating with very close relatives, the female-biased sex ratios which would maximize the number of offspring for a species as a whole would also be those that would maximize the fitness of individual genes.[5] That paper is to my mind the finest demonstration that the selfish gene framework is right. Hamilton's arguments make quantitative predictions about the sex ratios that will be seen with different levels of inbreeding, predictions which he (and subsequently others) showed were true for many species of insect across a bizarre range of natural histories. I reasoned that if sex ratio theory is the empirical bedrock of selfish genery, then we should apply it to infectious diseases. If that didn't work, harder things like disease virulence would surely be beyond us.

Most infectious disease agents do not have males and females. Malaria parasites do. In malignant human disease, the malaria cells that infect mosquitoes have what Ronald Ross called a semi-lunar shape.[6] These 'crescents' have sex in the mosquito, and it is possible to distinguish the male and female forms in our blood. Females dominate, in contrast to 1:1 sex ratios in the vast majority of free-living species. I reasoned that this was a Hamiltonian bias.

I discovered I wasn't the first to think this (Michael Ghiselin and John Pickering had thought of it earlier), but encouraged by colleagues, particularly Anne Keymer, David Walliker, and Paul Harvey, I pushed the idea harder. It turned out that we really could make quantitatively successful predictions. Karen Day and I measured malaria sex ratios in Papua New Guinea, and then, with Sean Nee, we used simple mathematical models to predict that at least 62 per cent of malaria zygotes in Papua New Guinea would be the result of mating among the same parasite clone.[7] At the time, conventional wisdom put the figure close to 0 per cent, though it had never been measured. Rereading our paper now, it is quite clear that we were worried about this wisdom, and I spent a great deal of the discussion describing why our estimate might be too high. I should have had more faith. Ric Paul and Karen Day subsequently used molecular genetic analyses to show that inbreeding rates in those malaria populations were in fact well in excess of 62 per cent.

This was the first scientific prediction I had ever made which had turned out to be right. Was it a fluke? Clearly we needed to do more, but with the best will in the world, no one was going to do lots of expensive molecular genetics just to test a fantasy of mine. So we needed a cheaper approach. I reasoned that sex ratios should be shaped by the rate at which people acquired new infections, and we could estimate that from the number of hosts that were infectious. Sean Nee and I formalized this mathematically, and then the search for data began. The data came in from collaborators over several years, mostly from populations of birds infected with malaria-like parasites. Each time we got data from a new population, I was shocked at the close fit between theory and observation. Eventually, we got data covering a large range of sex ratios and, staggeringly, *all* were as expected. Although it was published over ten years ago, I still consider that work to be my most philosophically satisfying.[8,9] From Chapter 9 of *The Selfish Gene*, it is possible successfully to predict previously unsuspected patterns for a group of organisms—and even a lifestyle—not featuring in *The Selfish Gene* or in almost any of the other work which flowed from it. This study relieved my physics-envy. For sure, not as impressive as predicting the existence of the planet Neptune in advance, but we evolutionary biologists can also make novel quantitative predictions that are right. And malaria is a damned sight more important to humanity than Neptune.

Of course, I failed to persuade anyone else that this was interesting. Evolutionary biologists already knew that sex ratio theory worked, and biomedical types simply didn't care: sex ratio does not affect how sick we get and, worse, our arguments involved non-intuitive theory and equations. These days, my collaborators and I are applying selfish genery to malarial virulence and infectiousness, and at least some biomedical people are interested. Our controversial prediction that some vaccines could prompt the evolution of nastier pathogens is just a small logical step from the sex ratio theory of Chapter 9; our discovery that selfish strains dominate infections shows that the kin selection and relatedness of Chapter 6 apply to malaria too.[10,11]

My sense is that *The Selfish Gene* had a huge impact among evolutionary biologists, ecologists, and behaviourists, recruiting people to these fields and helping to get right the thinking of the less mathematically inclined. But in biomedicine, the largest and most well-funded area of biology, selfish genery has had negligible impact. This is in part because evolution is largely absent from biomedical training, and also because evolutionary biologists have been slow to leave the comfortable natural histories of birds and insects for the jargon-laden natural history of medicine. But it is also a consequence of the overwhelming dominance of a reductionism in biomedicine (ironically a criticism once levelled at Dawkins). Explanation of disease virulence and infectiousness is usually sought in terms of molecular interactions, cell signalling, and so on. Mechanistic description is of course fantastically important and has yielded substantial insight and some clinical advances. However, such explanations are necessarily incomplete. To explain why something is like it is, we also need to ask about the evolutionary pressures. And this involves the thought processes laid out in *The Selfish Gene*.

Not thinking like this could even be dangerous. The conventional wisdom that infectious diseases evolve to be nice is hopelessly wrong. Evolution does not maximize longevity of an individual or the reproductive output of a species. If a virulent mutant competes more successfully with other parasites, that mutant will spread even if it is more likely to kill its host, its competitors, and itself. Had SARS persisted in the human population, would it evolve to be nastier or nicer? Would the intervention measures we would be throwing at it alter this evolution for better or for worse? Such questions are very rarely even asked, and we do not know the answers.

For malaria, there are two questions I want answered. Why are strains that produce more of the semi-lunar cells needed to infect mosquitoes not more common? Broadly speaking, more transmission stages beget more transmission, yet most malaria infections contain barely any. Something very interesting must be going on for selection to favour reproductive restraint.[12] Second, why are malaria parasites killing so few people? Our experimental work shows that virulent strains have a fitness advantage; something is stopping them spreading. The chance of an African being killed by a single dose of malaria parasites is less than 1 per cent. Why should mutant parasites running a 2 per cent risk not spread?

A selfish gene perspective naturally begs such questions and, as Dawkins showed so clearly thirty years ago, provides a means to answer them. For malaria, some selection pressure is keeping the lid on transmission and virulence. Ideally we would like to use public health measures to screw that lid down tighter. We certainly do not want inadvertently to loosen it.

ENDNOTES

1 Indeed, if kakapo are the result of Intelligent Design, the designer was not very far-sighted.

2 Summarized in R. M. Anderson and R. M. May, *Infectious Diseases of Humans: Dynamics and Control* (Oxford: Oxford University Press, 1991).

3 Summarized in P. Ewald, *Evolution of Infectious Disease* (Oxford: Oxford University Press, 1984).

4 J. R. Krebs and N. B. Davies (eds.), *Behavioural Ecology: An Evolutionary Approach* (Oxford: Blackwell, 1978; 2nd edn., 1984); *An Introduction to Behavioural Ecology* (Oxford: Blackwell, 1981). Like Dawkins, Krebs and Davies were animal behaviourists, with the

consequence that the field came to be called *behavioural* ecology. This is something of a misnomer; selfish genery extends way beyond behaviour.

5 W. D. Hamilton, 'Extraordinary sex ratios', *Science*, 156 (1967): 477–488.

6 Winner of the 1902 Nobel Prize for his verification of Manson's prediction that mosquitoes transmitted malaria. R. Ross, *Memoirs* (London: John Murray, 1923).

7 A. F. Read, A. Narara, S. Nee, A. E. Keymer, and K. P. Day, 'Gametocyte sex ratios as indirect measures of outcrossing rates in malaria', *Parasitology*, 104 (1992): 387–395.

8 A. F. Read, M. Anwar, D. Shutler, and S. Nee, 'Sex allocation and population structure in malaria and related parasitic protozoa', *Proceedings of the Royal Society of London Series B*, 260 (1995): 359–363.

9 Things have moved on since this work; for reviews aimed at (i) evolutionary biologists or (ii) parasitologists respectively, see (i) A. F. Read, T. G. Smith, S. Nee, and S. A. West, 'Sex allocation in microorganisms', in I. Hardy (ed.), *Sex Ratios: Concepts and Research Methods* (Cambridge: Cambridge University Press, 2002), 314–332, and (ii) S. A. West, S. E. Reece, and A. F. Read, 'Gametocyte sex ratios of malaria and related apicomplexan (protozoa) parasites', *Trends in Parasitology*, 17 (2001): 525–531.

10 The primary papers here are: (i) S. Gandon, M. J. Mackinnon, S. Nee, and A. F. Read, 'Imperfect vaccines and the evolution of pathogen virulence', *Nature*, 414 (2001): 751–756, and (ii) J. C. de Roode, R. Pansini, S. J. Cheesman, M. E. H. Helinski, S. Huijben, A. R. Wargo, A. S. Bell, B. H. K. Chan, D. Walliker, and A. F. Read, 'Virulence and competitive ability in genetically diverse malaria infections', *Proceedings of the National Academy of Sciences USA*, 102 (2005): 7624–7628.

11 For an overview of our virulence work, see (i) M. J. Mackinnon, and A. F. Read, 'Virulence in malaria: An evolutionary viewpoint', *Philosophical Transactions of the Royal Society of London Biological Sciences*, 359 (2004): 965–986, and (ii) A. F. Read, S. Gandon, S. Nee, and M. J. Mackinnon, 'The evolution of pathogen virulence in response to animal and public health interventions', in K. Dronamraj, (ed.), *Infectious Disease and Host-Pathogen Evolution* (Cambridge: Cambridge University Press, 2004), 265–292.

12 L. H. Taylor, and A. F. Read, 'Why so few transmission stages? Reproductive restraint by malaria parasites', *Parasitology Today*, 13 (1997): 135–140.

THE BATTLE OF THE
SEXES REVISITED

Helena Cronin

A FRUIT fly delivers his sperm in a toxic cocktail that consigns his consort to an early death. A female dung fly, caught in a scrum of eager suitors, is ignominiously drowned in a cowpat. A spider eats her mate in the very act of copulation. Dramas such as these epitomize the battle of the sexes. Or do they? What is that notorious battle really about? And how can we tell it from life's many other conflicts?

In *The Selfish Gene*, Richard Dawkins invites us to take a gene's-eye view of these and other Darwinian questions. Like Einstein's imagined ride on a beam of light, this is an invitation to journey into unreachable worlds for a clearer understanding of reality. It envisages the strategies of genes as they take their paths down the generations, over evolutionary time; the eloquent biographies speak to us of natural selection's design. Unlike most thought-experiments, this is not a solution to one particular problem but a way of perceiving the entire world of living things. And it is a method of great potency. It has immense explanatory power—not surprisingly, for it precisely captures the logic of natural selection's problem-solving; and thus it can generate testable hypotheses. It has remarkable predictive power, prising out telling but otherwise unappreciated evidence. It transforms our view of the familiar, turning into questions what had unthinkingly been regarded as answers. It reveals worlds undreamed of, alerting us to counterintuitive realms. And it dispels confusions, even the tenacious and the wilful.

I started in philosophy, where Darwinism was persistently maligned. Surveying the science, I rapidly concluded that the philosophers were profoundly wrong. *The Selfish Gene* became my staunchest guide. Here was a Darwinian world that was gene-centred, adaptationist; this had to be how natural selection worked. That, and *The Extended Phenotype*, introduced me to fundamental questions of evolutionary theory. And they taught me how, holding steadily to a gene-centred view, I could find the way through muddle; follow that gene and the rest will fall into place.

So how does this perspective contribute to our understanding of the battle of the sexes? Let's begin by reviewing the gene's-eye view of life. Genes are machines for turning out more genes; they are selfishly engaged in the dedicated pursuit of self-replication. The means by which genes propagate themselves are adaptations, devices that enable them to exploit whatever they can of the world's potential resources to survive, flourish, and replicate. Adaptations manifest themselves as the familiar design features of living things—tails, shells, petals, scents, the ability to glide on the breeze or beguile a mate. Differences in genes give rise to differences in adaptations. Natural selection acts on these differences and thereby on genes. Thus genes come to be represented in successive generations according to the success of their adaptations.

Among genes all is selfishness, every gene out for its own replication. But from conflict can come forth harmony; the very selfishness of genes can give rise to cooperation. For among the potential resources that genes can exploit is the potential for cooperation with other genes. And, if it pays to cooperate, natural selection will favour genes that do so. Thus selfish genes can come to be accomplished cooperators—selfish cooperators, pragmatic cooperators, but accomplished cooperators nonetheless. Their cooperation arises not in spite of but because of genetic selfishness.

And so, by working together, genes can enjoy the fruits of such magnificent feats of cooperation as the high-tech bio-chemical factory that is a cell; the orchestrated assembly line that is embryonic development; the elaborately equipped vehicles that are bodies. Each of these adaptations is incomparably more intricate, more effective than any gene could build alone. And each cooperative enterprise creates a platform for the next. Thus, from modest beginnings and from a foundation of implacable genetic selfishness, genes have evolved the means to transform the world's resources in ever more ingenious ways, proliferating adaptations of ever greater complexity and sophistication.

Cooperation is in principle possible wherever interests coincide. But interests are of course seldom, if ever, identical. So cooperation gives rise to potential conflict. It should, then, be no surprise if conflict is a persistent accompaniment to a cooperative enterprise —in particular if it occurs over the very resources that cooperation creates, the spoils of joint venture. Each cooperative endeavour generates new resources and thus new arenas of potential conflict.

So when conflict arises within a game of cooperation it arises at the margins, for that is where interests diverge. Even small margins can generate heated disagreement. This is because haggling over the margins is a zero-sum game, a game of 'your loss is my gain'. Think of a buyer and seller arguing over the price of a rug. A stranger to rug-markets would assume that they were arenas where strife runs deep. But such bargaining takes place within a game of cooperation—in this case, mutually beneficial trade.

Thus wherever we see cooperation, at whatever level, we should be prepared also to see conflict. This is crucial to bear in mind when we look at the battle of the sexes. For it is a battle played on the margins of a prodigiously successful cooperative enterprise—sexual reproduction. So the presence of conflict should not be interpreted as lack of cooperation; on the contrary, it will be the result of cooperation.

And so to sexual reproduction. It evolved as a solution to two troubles that cloning organisms are heir to: mutations and parasites. The problem with mutations—copying mistakes—is that they get faithfully copied along with the other genes, whether or not they work well together; the mistakes accumulate down the generations; and eventually they drive the lineage to extinction. The problem with parasites is that, once they have specialized in exploiting a particular host, that host's descendants can't shake off their sitting tenant and so are obliged to provide free board and lodging for as long as the host's lineage persists. Enter sexual reproduction, shuffling genes thoroughly with each generation, creating entirely novel collections of genes with each organism— and thereby both cleansing the gene pool of mutations that undermine the workings of organisms and presenting to parasites not the sitting target of clones but a target forever on the move.

With the advent of full-scale sexual reproduction—some 800 million years ago—came new tasks. A sexual organism must divide its total reproductive investment into two—competing for mates and caring for offspring; and whatever is spent on one is unavailable for the other.

From the very beginning of sexual reproduction there was an asymmetry in investment, a sex difference—one sex specializing slightly more in competing for mates and the other slightly more in caring for offspring. This arose because there were two distinct groups of genes and they got into the next generation in different ways. Genes were housed (as they are today) in two parts of the cell—the vast majority in the nucleus and a few outside it, mainly in the mitochondria, the cell's powerhouse. Nuclear genes went into both kinds of sex cells, half of them in each cell. But, because two sets of mitochondria would lead to disruptive conflict over which should be the powerhouse and which would be surplus to requirements, they were allowed into only one kind of sex cell. So one sex cell started out larger and with more essential resources than the other. And thus began the great divide into fat, resource-laden eggs, already investing in providing for offspring, and slim, streamlined sperm, already competing for that investment.

Once that divergence had opened up, evolutionary logic dictated that it became self-reinforcing. If you specialize in competing, you gain most selective advantage by putting more into competing; and the same for caring. And so the divergence widened over evolutionary time, with natural selection proliferating and amplifying the differences, down the generations, in every sexually reproducing species that has ever existed.

Thus, from such inauspicious beginnings, from this slight initial asymmetry, flow all the characteristic differences between males and females throughout the living world, differences that pervade what constitutes being male or female. Indeed, for evolutionary biologists, downloading mitochondria into future generations or offloading them into genetic oblivion is the fundamental difference between females and males.

What do those sex differences typically look like? A female possesses a scarce and valuable resource: her eggs and the investment that goes with them. She needs to be judicious about how she ties up this precious commodity. So she goes for quality. Which qualities? Good genes and good resources, particularly food, shelter, and protection for herself and her offspring. Meanwhile, a male's reproductive success is limited only by the number of matings that he can get. So he goes for quantity. Thus males compete with one another to provide what females want. They display their quality with costly, elaborate ornaments; and they strive to get resources and to hold on to them.

So males, far more than females, compete to be the biggest, brightest, brashest, and best; they are larger and stronger; they take more risks and fight more; they care more about status and power; they are more promiscuous and do less child-care; and they expend vast quantities of time, energy, and resources just strutting their stuff—singing, dancing, roaring; flaunting colours and iridescence; displaying tails and horns . . . adaptations in glorious profusion.

Now that we know what the two sexes want, we can see how and why males and females come into conflict. He wants to mate more often than she does; she wants to be choosier than he does. And she wants to extract more parental investment from him than he wants to give; he—vice versa.

Conflicts over mate choice have led males into advertising and deception, stealth and force—and females into counteradaptations ranging from lie-detectors to anti-clamping devices. The escalation of extravagant male advertisement driven by female scrutiny is famously exemplified by the peacock's tail, a story told in the second edition of *The Selfish Gene*. Male force and a woman's fight to choose are vividly illustrated by the female reproductive tract of a multitude of insect species. She has evolved Byzantine arrangements of chambers and corridors, lumber rooms and labyrinths for storing sperm to be used when she chooses. And he has evolved an armature of scoops and toxins, hooks and horns to oust the sperm of other males, and an array of glues and plugs to prevent her or would-be later lovers from ousting his. As for adaptations for conflicts over parental investment—tactics for monopolizing or increasing the other's investment and minimizing one's own—this story, too, is told in *The Selfish Gene*. And any sample of recent papers is striking testimony to how much we now know about the multitude of inventive arguments to be had over who looks after the children: 'why don't male mammals lactate?'; 'mother/father differences in response to infant crying'; female starlings 'increase their copulation rates when faced with the risk of polygyny'; 'female-coerced monogamy in burying beetles'.

Let's now parse a case involving both mate choice and parental care, disentangling conflict from cooperation and the strategies of genes from the behaviour of the males and females that house them. Picture a pair of titi monkeys, husband and wife in close embrace, their tails entwined, in sleep cuddled together, when awake always close, preferring one another's company above that of all others—and with so little sex difference that they look more like twins than spouses. What is the recipe for their successful marriage? It's that he takes on a huge burden of child-care; he's an authentic New Man. During the baby's first fortnight, apart from its feeding times, he carries it constantly; and his care continues to be so assiduous that the baby is more distressed at his disappearance than at its mother's. And as a wealth of evidence shows—including cross-species comparisons and behavioural and hormonal measures—their parental division of labour reflects cooperation among their 'parental-investment' genes. But continue to trace the interests of their respective genes and we find that what might appear to be the very epitome of their happy pact—the entwined tails, the constant cuddles—reflects instead genetic conflict. For they are mate-guarding: an adaptation that reflects an evolutionary legacy of less than perfect monogamy (which continues still in titis, albeit rarely). He is protecting himself from lavishing his investment on another male's offspring; she, however, could benefit if her putative lover had superior genes. She is protecting herself from the danger of losing his paternal investment; he, however, could benefit from additional matings elsewhere. So their conflict is over mate choice. And it is engendered by the very resource, parental investment, that their cooperation has created. What joins them together has also— among their genes—put them asunder.

Now to a case of parental conflict more arcane and more wondrous. It takes us to a prediction that, when *The Selfish Gene* was published, was a mere twinkle in the gene's-eye view; although not spelt out explicitly in the book, these tactics of genes brimmed from the logic. To begin, let's remind ourselves of how to enter into the gene's view. It involves following the trajectory of genes, tracing the careers of the strategies by which they replicate themselves. We must think, then, not about a single encounter between a male and a female, nor about their success over a mating season, nor even about all the encounters during their lifetimes but about the average success of a gene over all its numerous instantiations, in different individuals, over many generations, down evolutionary time. And we must bear in mind that, in sexually reproducing species, most genes move back and forth between male and female bodies, spending 50 per cent of their time in each; so genes responsible for sex differences will alter their tactics to fit with their current neighbours, following the rule: 'If in a male body, do this; if in a female body, do that'.

Now to the example. 'Don't argue in front of the children!' goes the standard advice. But what about arguing *in* the children? Consider this gene-centred thought-experiment. Imagine a gene travelling down the generations in successive offspring and having different interests depending on whether it has just come from the father or the mother. For the interests of paternally- and maternally-derived genes will indeed diverge, particularly if the species cannot boast a history of unalloyed female monogamy (which no species can). And one divergence arises because maternally-derived genes will be in all the future offspring of that mother whereas paternally-derived genes might be in none. Now imagine, too, that the mother provides generous nutrients inside her body for the fertilized egg; so the entire burden of care is on her and the father is absolved. That immediately creates an arena for conflict over maternal provisions. The maternally-derived genes want the mother to provide not only for themselves but also for copies of themselves in her future offspring; but the paternally-derived genes, having no guaranteed interest in future offspring, want to exploit the mother's body beyond that 'fair share'.

Impeccable gene-logic. But thirty years ago the facts looked unsympathetic; orthodox thinking was that genes couldn't possibly know which parent they came from. It turns out, however, that they can—it's called 'genomic imprinting'—and, what's more, that they behave precisely as predicted. Molecular biologists, after initial astonishment, are now accustomed to discovering such genes regularly in both mammals and flowering plants.

Take a mouse species in which the mother is near-monogamous; all the offspring in a litter have the same father and there's an 80 per cent chance that he'll father the next litter. And take also a closely related species in which there is a plethora of fathers within litters and from litter to litter. Cross a 'multiple-paternity'-species male with a 'monogamous-mother'-species female. Result: offspring larger than normal. This is because the father-derived genes wrest all that they can from the mother and the mother-derived genes are unaccustomed to putting up much resistance. Cross the other way—a 'monogamous-mother' male with a 'multiple-paternity' female. Result: offspring smaller than normal. The mother-derived genes are taking their miserly 'fair share' without the usual top-up from the father-derived genes. Compare the results: dramatic differences in birth weight. The genes' battle is a tug-of-war, each side assuming resistance from the other, the tugs escalating in strength over the generations. The experiment forces one side to let go; and the entire game comes tumbling down. Even without laboratory intervention one side sometimes lets go. This explains, for example, much about the typical pathologies of human pregnancy, otherwise so baffling in a well-honed adaptation. Note that the tug-of-war is a marginal conflict arising in a cooperative game. The parents have a vast area of overlapping interests, for both want the offspring to develop normally. But conflict over the size of the mother's investment has triggered an evolutionary arms race and that has settled into the tug-of-war. When both sides tug normally, the result is a compromise growth rate and a normal baby. But either side letting go spells disaster for all—for genes in the embryo and in the parents.

Why does this particular battle of the sexes occur in mammals and flowering plants? Because, in both, the mother provides nutrients after fertilization. Maternal genes have hit on the same strategies for solving a 'caring' problem; and so they have generated the same arena for conflict, the father trying to get just that bit more than the mother's rationing stipulates. Thus the gene's-eye view has revealed hitherto unexplored commonalities across vast taxonomic divides—genes that have converged on the same strategic solutions and faced the same resultant conflicts.

And the embryo is just one such battleground. The same logic could get to work wherever the interests of relatives on the mother's side and the father's side diverge. So watch out for it in closely related groups; it could shed unexpected light on family rows.

Having seen what the battle of the sexes is, we are now in a position to see what it is not—to identify cases that might look persuasively like such a battle but aren't. For the battle of the sexes, there must be a conflict of interest between 'male' and 'female' genes (genes implementing male and female strategies)—not just a tussle between male and female bodies. And the conflict must be over mate choice or parental investment.

We can begin by clearing away that dead dung fly. Although she was drowned in a scrum of over-eager males, neither she nor her sisters who are often cited as victims of 'male-induced harm'—insect, bird, and reptile, drowned, crushed, and suffocated —are casualties of the battle between the sexes. Such battles are within one sex, males against males; the females are civilians caught in crossfire. There is no adaptation for courting by drowning; indeed, the winner's prize is a Pyrrhic victory.

The corpse of the poisoned fruit fly is, by and large, also collateral damage. The toxins in her suitor's seminal fluids are a cocktail that search-and-destroy the sperm of previous suitors; trigger her egg production; and slip her an anti-aphrodisiac intended to last until her eggs are fertilized. Insofar as his genes attempt to control when and by whom she is fertilized, she is certainly a victim of the battle of the sexes. But the main cause of her early death is a conflict between males who use her body as their battleground—sperm competition so fierce that the drugs take an unintended toll. Indeed, remove sperm competition from their mating and give him a vested interest in her as a long-term baby-machine—as was done in an experiment that imposed monogamy on both of them—and, within a few generations, his seminal fluid becomes increasingly less toxic.

Thus reducing the poison when monogamous is an adaptation; poisoning her when promiscuous is not. Jesuitical as such distinctions might seem to a drowned or poisoned bride, for a gene-centred analysis it is crucial to ask whether the adaptation is a weapon in the battle of the sexes and, if so, whether designed to harm her. Admittedly, such scrupulous partitioning of sexual conflicts into male–female and male–male is not always germane. But, before we attribute a conflict to the battle of the sexes, we should bear in mind that both selective forces could be at work.

Now to an example of harm to one of the partners that is not collateral damage and indeed, contrary to dramatic appearances, is probably not a result of conflict at all. The female redback spider consumes her mate while they copulate. But genetic conflict? No. He is investing in his offspring. Some males feed their young with bodies that they have caught; he cuts out the middle-man and offers up his own. In a virtuoso delivery of fast food, he flips a somersault that lands him neatly in her mouth. Such suicidal behaviour usually evolves when males have little chance of finding another mate, which is indeed the case for him. However, he might also have a further agenda: sperm competition. By preoccupying her with food, the disappearing male can prolong copulation—far longer than needed to transfer sperm but just right for adding a poisoned parting shot for rival sperm. If so, he has moved from giving paternal care to circumventing her mate choice: from cooperation to the battlefield of the sexes.

Female cannibalism can be against the lover's will but even so not necessarily a battle of the sexes. The female praying mantis is notorious for her voracious sexual cannibalism. About one-third of her matings are also meals, consumed during or after copulation. The male has evolved elaborate adaptations against being eaten; he approaches stealthily and pounces suddenly on her back while she is distracted, a finely orchestrated suite of moves. So there is certainly conflict in their encounter. But whereas on a poor diet she devours about three-quarters of her mates, on a rich diet she downs less than a quarter. So perhaps her adaptation is not to eat her lover but just to eat, not gratefully to receive paternal investment but just to have lunch. If, to her, he is indeed meat not mate, then their battle is not that of the sexes but that of predator and prey.

Finally, an apparently flagrant case, the very cutting edge of sexual conflict: females castrating males, a practice rife among flowering plants. But it turns out not to be a battle between the sexes although, among organisms, the male parts of the plant wither while the female parts on that very same plant flourish. The battle is between mitochondrial and nuclear genes. Remember that, from the beginning of sexual reproduction, mitochondria have gone only into eggs, never into sperm; so they have travelled solely through the female line. Thus—although they replicate by cloning—they rely on the sexual reproduction of the body that they are in to exit, in fertilized eggs, to the next generation. However, whereas when they exit into a daughter they are again in a vehicle to the next generation, when they exit into a son the vehicle is their hearse, for they cannot get into sperm. And so mitochondria in male parts in those flowering plants, rather than helping in the sperm-making factory as they should do, subvert the assembly line so that the sperm aren't viable. Such unruly behaviour is strongly predicted by gene-centred theory. Genes that travel as a band down the generations through only one sex would not want to get into the other sex; and so they would want to gang up against it. The ancient arrangement with mitochondria was a recipe for conflict. For plant species, it becomes a conflict about whether to reproduce sexually at all.

We are now ready to survey afresh the battlefield of the sexes. With *The Selfish Gene* as our guide, we can compile a gene-centred inventory of any participants that we come across. There are the principal antagonists: 'male' and 'female' genes in conflict and the resulting multifarious adaptations in male and female creatures. There are the victors and victims that we would other-wise have missed: the agents of genomic imprinting, visiting the feuds of the previous generation upon the children. There are the subversive opportunists: mitochondria waging their own private war against nuclear genes over their mitochondrial mausoleum, the male parts of plants. And there are the innocent bystanders: victims of collateral damage from other's battles, such as drowned and poisoned females.

So the banner of the battle of the sexes should depict not a spider's quietus in his lover's jaws nor the withered anthers of a flower; for they represent other concerns of genes. Nor need it depict struggle or pain, injury or death. It could instead portray the dazzling beauty of the peacock's tail, the deep intimacy of the titis' embrace, the finely poised equilibrium that builds a newborn baby. Such are the ways of genes that even their conflicts—'male' against 'female' genes—can appear deceptively to us as harmony and beauty in their bearers.

A wider survey of the battlefield reminds us that, however salient the conflicts, however fierce the battles, they are but mar-ginal within the vast and intricate game of cooperation that is sexual reproduction. And this cooperation reminds us—for there are those that need reminding still—that selfish genes do cooperate with one another. Indeed, as *The Ancestor's Tale* emphasizes, even with the gene-shuffling of sexual reproduction, the genes of a species become good cooperators.

[G]iven sex, . . . genes are continually being tried out against different genetic backgrounds. In every generation, a gene is shuffled into a new team of companions, meaning the other genes with which it shares a body . . . Genes that are habitually good companions . . . tend to be in winning teams—meaning successful individual bodies that pass them on to offspring . . . But in the long term, the set of genes with which it [a gene] has to co-operate are all the genes of the gene pool, for they are the ones that it repeatedly encounters as it hops from body to body down the generations.[1]

And, finally, moving beyond the battlefield, *The Selfish Gene* reminds us that, although the bearers in which genes move down the generations are engaged in enterprises of apparently great pith and moment, a genes'-eye perspective takes our understanding far beyond these transitory players, opening up an immense reach down time, along the deep flow of immortal genes.

ENDNOTE

1 Richard Dawkins, *The Ancestor's Tale* (London: Weidenfeld & Nicolson, 2004), 359.

RICHARD DAWKINS: INTELLECTUAL PLUMBER—AND MORE

John Krebs

T HE other week I was enduring one of those dinner parties at which the strength with which opinions are held is matched only by the weakness of their foundation.

When the conversation turned, as it all too often does, to the possible health benefits of Organic Food, I was asked for a view in my capacity as the recently demitted Chairman of the UK Food Standards Agency. As I embarked on a brief, but pithy, exegesis of the scientific evidence, I was cut short in the middle of my first paragraph by the braying voice of the woman seated opposite me who interjected: '*But we don't believe in science in our family*'.

Suppressing my immediate inclination towards some form of physical response, I diverted my thoughts, as I often have before in similar situations, to the idea that there really should be an 'Emergency Dawkins Service'. When you have a leak in your pipes you call in a plumber; when you have leaky reasoning you call in Dawkins, the Handy Intellectual Plumber. One, or at most two, sessions would have sorted out my fellow diner.

As Richard expressed it so eloquently in *River Out of Eden*:

Show me a cultural relativist at 30,000 feet and I'll show you a hypocrite. Airplanes built according to scientific principles work. They stay aloft and they get you to your chosen destination. Airplanes built to tribal or mythological specifications, such as the dummy planes of cargo cults in jungle clearings or the beeswaxed wings of Icarus, don't.

But Richard is more than a plumber who can fix leaky intellects.

When he first started to write about evolution, his academic colleagues sought to place him on a spectrum that has, at one end, 'vulgar popularizer' and, at the other, 'scientist's scientist—a creator of pure new knowledge'. However, Richard defies classification on this spectrum. What he does, and better than anyone else, is to reanalyse or reinterpret the findings of others with such excoriating rigour, depth, and clarity that he uncovers new ideas and ways of thinking. More often than not, the author of the original research will gain new insights into the significance of their findings as a result of Richard's reanalysis and interpretation.

Richard's reinterpretation of animal communication,[1] in a paper on which I was (very much) a junior author makes the point. During the previous forty years, ethologists had established the canon that animal communication has evolved for the mutual benefit of signallers and receivers. Amongst the subscribers to this view were all the major ethologists of the mid-twentieth century, including the great Niko Tinbergen ('One party—the actor—emits a signal, to which the other party—the reactor—responds in such a way that the welfare of the species is promoted').[2] Even the awesomely intelligent J. M. Cullen (mentor for both Richard and myself) accepted the 'information for mutual benefit' view of animal signalling in his masterly 1966 review of ritualization.[3]

Anyone familiar with 'selfish gene' thinking will immediately spot the problem. The view that communication evolves for mutual benefit is essentially an argument based on the premise that natural selection works for the good of the group or species, rather than the good of the gene. It is especially remarkable that, in spite of this implicit underpinning assumption, many of the major proponents of the 'information' view of animal communication were avid, neo-Darwinian, individual selectionists. It took Richard's relentless, uncompromising, and surgical application of neo-Darwinian thinking to expose the logic of animal communication with coruscating clarity.

In turning forty years of work by the leaders of the field on its head, Richard redefined animal communication in the following way:

Natural selection favours individuals who successfully manipulate the behaviour of other individuals, whether or not this is to the advantage of the manipulated individuals. Of course selection will also work on individuals to make them resist manipulation if this is to their disadvantage . . . Actors do sometimes succeed in subverting the nervous systems of reactors, and the adaptations to do this are the phenomena we see as animal communication.[4]

This reinterpretation of animal communication immediately became the new canon for studies of signalling. The idea of manipulation had been foreshadowed for specific aspects of communication in other papers,[5,6] but Richard's paper generalized it, put it on the map, and opened up a new window through which to view old ideas, such as the ritualization of signals. Ritualization is the elaboration of signals during the course of evolution from their simpler ancestral forms. It had been seen, since the work of J. S. Huxley in the early part of the twentieth century, as an expression of selection to improve the efficacy of information transfer. For instance, in the duck family, ancestral preening movements have become exaggerated, stereotyped, and repetitive as part of male courtship in several species. By observing the varying degrees of divergence from the ancestral movement among present-day species, we can infer a hypothesized path by which it became ritualized. Richard's reinterpretation was that ritualization was the expression of an arms race between signaller and receiver nervous systems.

The new paradigm not only resulted in a radical reinterpretation of all the old data, but also enabled the next generation of behavioural ecologists to develop and expand the idea with theory and experiment in contexts such as mate choice, territoriality, and parent–offspring communication.[7] In a closely related and complementary strand of thinking, the highly original Israeli scientist, Amotz Zahavi[8] had suggested in the mid-1970s that as a result of the evolutionary arms race between signallers and receivers only honest signals would persist, and that honesty could only be guaranteed if the signal has a cost to the signaller. Zahavi's hypothesis would, for example, postulate that the bright colours of many male birds are honest signals of individual quality, either because the colours themselves are costly to produce, or because only a high-quality male can afford the encumbrance of showy plumage that renders him more conspicuous to predators. This concept of the Handicap Principle is the other side of the coin of manipulation.

I have dwelt on this example to illustrate the general point that Richard is much more than a summarizer and popularizer of other people's ideas: he is a genuine creator of new science.

Richard is, for most readers, known only for his writings on evolution, so it is worth touching on some of his other contributions. Back in the 1960s and early 1970s, when PDP 8 and PDP 11 computers the size of a room had far less capacity than does a hand calculator today, and when you had to store your programs and data on reams of paper tape or enormous stacks of punch cards, Richard was right at the forefront of their use in recording and analysing behavioural data. He dragged us fellow members of the Animal Behaviour Group at Oxford into the computer age, teaching us how to write programs in machine code. He also invented the 'Dawkins Organ', an early event recorder that enabled one to record behavioural data as tones on a continuously running magnetic tape, to be sub- sequently decoded by one of those room-sized PDP 11s. I was one of the lucky first beneficiaries of this step change in data processing technology when I worked on Great Blue Herons at the University of British Columbia in the early 1970s.

Going even further back, to his D.Phil. thesis, Richard analysed the pecking behaviour of day-old chicks and the ontogeny of the cues by which they recognized a solid object. I recall reading his thesis as a potential model when I was writing mine a few years later. I knew I was reading the work of a Master when I saw on the first page of the Methods that 'The chicks were tested in Paris'. A misprint perhaps, but it certainly conveyed the impression of a certain style!

Richard has spent most of his career at Oxford, but for two years in the late 1960s he was at Berkeley, during the time of hippies, student riots, and revolution. On his return he recounted how one day, while he was walking down Haight Ashbury, no doubt on his way to a bookstore and wearing empire-building shorts and well-trimmed hair, a car full of gawping tourists drove slowly past him, and a child inside was heard exclaiming '*Hey Maw, it's one of the weirdos!*'

ENDNOTES

1   R. Dawkins and J. R. Krebs, 'Animal Signals: Information or manipulation', in J. R. Krebs and N. B. Davies (eds.), *Behavioural Ecology: An Evolutionary Approach* (Oxford: Blackwell, 1st edn., 1978), 282–309.

2   N. Tinbergen, 'The evolution of signalling devices', in W. Etkin (ed.), *Social Behavior and Organisation Among Vertebrates* (Chicago and London: University of Chicago Press, 1964), 206–230.

3   J. M. Cullen, 'Reduction of ambiguity through ritualisation', *Philosophical Transactions of the Royal Society B*, 251 (1966): 363–374.

4   R. Dawkins and J. R. Krebs, 'Animal Signals: Information or manipulation', in J. R. Krebs and N. B. Davies (eds.), *Behavioural Ecology: An Evolutionary Approach* (Oxford: Blackwell, 1st edn., 1978), 282–309.

5   R. L. Trivers, 'Parent–offspring conflict', *American Zoology*, 14 (1964): 249–264.

6   E. L. Charnov and J. R. Krebs, 'The evolution of alarm calls: Altruism or manipulation?', *American Nature*, 110 (1975): 247–259.

7   D. Harper, 'Communication', in J. R. Krebs and N. B. Davies (eds.), *Behavioural Ecology: An Evolutionary Approach* (Oxford: Blackwell, 3rd edn., 1991), 374–397.

8   A. Zahavi, 'Mate selection—selection for a handicap', *Journal of Theoretical Biology*, 53 (1975): 205–214.

WHAT IS A PUMA?

Michael Hansell

A PANTHER is on the loose in the west of Scotland. Sightings suggest it is moving towards Glasgow. There is no photo, no DNA extracted from a hair found on a garden fence, but it is out there. There are others down in England.

In the 1960s the puma, not the panther, was the wild beast of the moment and in the summer of 1965 a bunch of Oxford zoology postgraduates and their friends set off on weekend safari in search of conclusive evidence of the 'Surrey puma'. Among the group were Richard Dawkins and myself.

The best evidence we came up with was some large scratch marks. A puma or, more likely, the foot flourish of a large dog satisfied after just relieving itself? Either way, the claw marks were not large enough for an accompanying press photographer who, with a few strokes of a stick, gave them some added terror. I have a recollection of talking to Richard about what would qualify as being a puma—a full profile, a face, a tail? Could it be said, I wondered, that the scratch marks were a puma I don't recall our conclusion but will return to the question.

In Oxford I was studying the organization of the behaviour by which caddis-fly larvae build themselves protective, portable homes, but was interested in animal building behaviour generally. Richard was at the time studying decision-making in chicks and was interested in all kinds of things but, a few years later, it was about caddis case building that he contacted me. Was there evidence for the genetic determination of some architectural feature of a caddis case? Richard was working on *The Extended Phenotype*, which made its appearance in 1982.

The Extended Phenotype is a development of the core idea established in *The Selfish Gene* that natural selection operates at the level of the gene rather than the organism. The subtitle of the 1982 edition was in fact *The gene as the unit of selection*. An animal-built structure happens to be a particularly powerful way of illustrating this. A whale and its tail are inseparable, so we feel a need to argue that it is the animal that is selected for the power of its tail. A portable sand grain tube is clearly not a caddis larva, forcing us to acknowledge that it is the expression of genes within the larva for building behaviour. The sand grain case of a certain species has one or more sticks, longer than the case itself, attached to its side. The sticks, it is now known, are a protection against fish predators. We do not know the genetic basis of this case character but, for the sake of simplicity, let us imagine that it is determined by a single gene location at which there can be only wo forms of the gene, or two *alleles*, to use genetic terminology. One allele (S) we will say is for attaching a stick, the other (s) for not attaching a stick. Like most organisms including ourselves, caddis larvae have two sets of chromosomes, one from each parent, giving them a pair of genes at the 'stick attaching' location. These could be any of the *genotypes SS, Ss*, or *ss* but if, let's say the expression of S (i.e. performance of stick attaching behaviour) is dominant over s, both SS and Ss will express the *phenotype* (the physical appearance of completed case) of an attached stick and only ss will not. Where fish predators abound, any stickless cases will disappear fast, along with the ss genotypes that created them. In a population that breeds freely, the s allele will become far less frequent than the S allele. However, in a habitat free of fish predation, stickless cases are not penalized and the cost of dragging around a heavy stick might in fact penalize the S relative to the s allele.

Caddis cases, bird and wasp nests have been the bread and butter of my research for over forty years. Consequently, I see *The Extended Phenotype* as illuminating the process of evolution in the field of animal building behaviour, and in this essay I want to illustrate the importance of the book through that. The danger, of course, is that in doing so the importance of Dawkins' innovative perspective on evolution will appear largely limited to animal building behaviour. Indeed, the very fact that I have been asked to write this essay shows that it is with my chosen field that the book has been particularly associated, whereas the subject of animal artefacts occupies but one of its fourteen chapters. The truth of course is that, as with all Richard Dawkins' other writing, the purpose of *The Extended Phenotype* is bolder and much broader in its implications. In his own words it is 'to free the selfish gene from the individual organism which has been its conceptual prison'. A caddis case just happens to be a perfect way to visualize that.

During the Carboniferous Period, about 380 million years ago, there was indeed a beast roaming what is now the west of Scotland; it was a two-metre centipede-like creature (*Arthropleura*). We know this from fossil footprints preserved in sandstone on the Isle of Arran. Are these footprints an extended phenotype? If so, it might appear that any grass blade bent by a passing beetle would deserve the name but consequently dilute the concept. *The Extended Phenotype* addresses this worry. The key question is, does natural selection act upon these footprints? This would be true if for example some Carboniferous predatory amphibian had stalked *Arthropleura* by following its footprints. Of course we have no idea, but it seems reasonable to argue that the making of these footprints had no effect on the survival of footprint makers or, as I should say, on the frequency of alleles linked to footprint creation. Consequently, there seems no merit in calling the footprints an extended phenotype. The Surrey puma question can now be answered: the scratch marks (even allowing for media misrepresentation) were not themselves a puma or for that matter a dog. They were certainly the phenotypic expression of genes for scratching behaviour. Yet, in the absence of any evidence that they affect the survival of scratch-generating genes, they do not deserve the description of extended phenotype.

For the 1999 edition of *The Extended Phenotype*, the subtitle was changed; it now reads *The long reach of the gene*. The case of the caddis is an example of this too, as it shows how genes sitting in the larva are responsible for producing a structure detached from it. But an equally eloquent illustration is the way in which a parasite can manipulate the behaviour of its host. This forms the subject of another chapter of the book. The freshwater 'shrimp' *Gammarus* normally swims away from the light to remain concealed from predators but, when infected with a certain type of parasitic worm, it swims towards the light, probably making it more liable to predation by ducks which are the final host in the life cycle of the worm. Here, the extended phenotype is expressed by one individual (the shrimp host) through the influence of genes contained in another individual (the parasitic worm). As it happens, we can extend the reach of the gene one stage further with an example in which the extended phenotype is a structure built by the host as an expression of genes represented in the parasite.

The host in this case is the caterpillar of the large white (or 'cabbage white') butterfly which, on reaching maturity, would normally spin a platform of silk from which it can hang after becoming a pupa or chrysalis. A caterpillar might not be so fortunate. It could have twenty or so eggs injected into it by a minute braconid wasp. These will hatch into parasitic larvae, consuming the caterpillar's internal organs before forcing their way through its body wall, spinning cocoons in a cluster on the leaf, and pupating. Remarkably and horrifically, the parasites leave enough of the tissues of the doomed caterpillar that it does not immediately die. Instead, it spins a tight silk web over the parasite cocoons, securing them to the leaf, and remains crouched over them, driving away any hyper-parasitic insects (parasites of parasites) with vigorous thrashing movements of its body. Manipulated by the parasites, the caterpillar has become their protector, even as it dies.

Notice here that the extended phenotype is the collective outcome of genes dispersed among several parasites, emphasizing the difficulty of regarding the individual as the unit of selection. For me this collective outcome also raises quite puzzling questions about how cooperatively-built structures, such as social insect nests, might evolve. This question is addressed with great insight by Richard Dawkins in *The Extended Phenotype* but, as a preliminary to describing this, it may help to know something about the village of Lavenham in Suffolk. Tourists flock to the heart of the town to see its fine sixteenth-century timber framed houses.

Towards the edge of the town are nineteenth-century brick houses. Clearly the houses in the two locations are made of different materials, timber frame being replaced by brick as the local supply of large timbers became scarcer and transport costs for bricks became lower. I'm going to call this change one of *technology*. These Lavenham houses are also different in style, partly dictated by the different materials, it's true, but a nineteenth-century brick house looks different to a brick house of the twenty-first or sixteenth-century. This change is one of *design*. In animal- as well as in human-built structures it is possible for design and technology to change independent of one another. What still startles me on rereading *The Extended Phenotype* is how clearly Richard Dawkins already understood this in 1982 and realized that it had important implications for nest building by social insects.

A mature mound of *Macrotermes* termites may contain three or four million individuals but they are all the progeny of a founding pair, the Queen and King. Since the mechanism of inheritance in termites is just like ours, these brother and sister workers will show the same range of differences in behaviour (that is, differences in their phenotype) as do human siblings. So how do these brothers and sisters agree what to build? Richard Dawkins recognizes in *The Extended Phenotype* that the solutions might be different depending on whether the problem is one of design or of technology.

In the case of the technology problem, he imagines a hypothetical termite in which choosing dark or light mud as a building material is determined by a pair of alleles at one gene location, and that selecting dark (D) is dominant over selecting light (d). If both the Queen and King are of genotype Dd, they will both produce equal numbers of D and d gametes (eggs or sperm). Pairing randomly, these will produce equal numbers of offspring with DD, Dd, dD and dd genotypes, producing a ratio of behavioural phenotypes in the workforce of 3:1 in favour of choosing dark mud. As the nest is built up from many thousands of mud loads contributed by a huge workforce, Dawkins' conclusion is that the mound will be a blend of dark and light in these proportions. What you might call a solution through proportional representation. Evidence from my own work suggests that such a system might exist.

In the 1970s I began to study a group of social wasps found in the rainforests of South East Asia. There are perhaps a score of species of these dainty and docile insects, which are collectively known as the hover wasps. One of the reasons I was attracted to hover wasps was that their nest designs were known to differ markedly between species. On direct acquaintance with them I discovered that the nest materials showed similar variation, exhibiting the full range of possibilities from totally mineral (let's just call it mud) to totally organic (a pulp of rotted vegetation). Some variation in the proportions of materials is even shown between nests of the same species. Unfortunately, we still have no idea if different individuals within the same colony are contributing more of one or other kind of material but, if we suppose for a moment that they are, how would decision making change in the face of selection pressure favouring more or less mud?

Imagine the wasp larvae being raised in nests that are parasitized by insects that pierce the nest wall with a hypodermic needle-like tube or ovipositor to lay their eggs in the wasp larvae. The mineral content of the mud walls protects against this so, where the threat of this parasite is high, nests with higher mud content produce more new nest founders. But mud is heavy and restricts nest attachment to rocky overhangs; lighter organic material nests can be dispersed through the forest, so reducing nest crowding and competition. Where insect parasites are rare therefore, nests with higher organic content will out-reproduce mineral ones. None of this has been demonstrated but the point is that a system of proportional representation in the workforce would generate differences between nests, and selection pressure sustained over generations would then result in the evolution of nest material technology.

In matters of nest design Richard Dawkins realized that it is harder to predict how disputes could be resolved but suggests, for a hypothetical termite colony, resolution through a voting system in which the majority will prevails. This of course requires the termites to convey their wishes to one another. This expectation is not unreasonable, he reminds us, because this is how honeybee swarms decide which of two rival nest sites is to be chosen. But in honeybees, nest relocation is a rare event whereas building in a termite colony is essentially constant and dispersed throughout the mound. Here, voting to resolve design disputes could be slow and complicated.

In the 1950s and 1960s, the French biologist Pierre-Paul Grassé conducted a series of classic experiments which showed how termites are stimulated to build a wall around their queen, whose body is a huge sausage, bloated with eggs. She emits a chemical signal (pheromone) from her whole body surface, which creates a gradient of the chemical in the air around her, with a high concentration near her, and lower concentration with increasing distance. The termites build the wall in a particular position determined by a critical value of the pheromone concentration. Well, that is the conventional story, but what happens where the workers differ in the genes they carry for the threshold values? In this case, if workers with a lower than the average threshold value build a wall a certain distance from the queen, workers with a higher than average threshold value might add material to its inside surface, so producing a thicker than necessary wall. It seems possible that high threshold individuals would succeed in doing this, even if in the minority, unless there was some mechanism that prevented it. We lack the experimental evidence to answer this question of worker disagreement, and it is possible that the evolutionary process has as yet not come up with a solution, leaving termite mounds quite inefficiently designed. However, some interesting insights on collective building have recently come from computer simulations in which groups of virtual builders build nests in an imagined three-dimensional space. These model systems are called *lattice swarms*.

In a lattice swarm model, groups of virtual colony members move randomly in a three-dimensional space or lattice, adding standard building bricks shaped like the hexagonal cells of wasp nests when stimulated to do so by specific local architectural configurations. These 'builders' are simple creatures, endowed with limited behavioural and sensory capacities; they do not communicate with one another, have no plan or blueprint, and no memory, attributes similar to, but a bit more limited than, those known in wasps. The question is, can a colony of such creatures create anything like a nest?

The answer, surprisingly, turns out to be yes. Some sets of rules (algorithms) produce comb-like designs which are orderly and modular, that is, showing regularity in the shape and arrangement of combs—a very characteristic feature of nest architecture across wasp species. These are so-called *coordinated* algorithms. In addition, any two architectures that resemble one another are found to be generated by similar algorithms. Coordinated algorithms, however, turn out to be quite rare; the majority prove to be *uncoordinated*. Each of these produces a more irregular architecture which is different every time the programme is run.

These lattice swarm models were not designed to test what would happen if some colony members used a different set of building rules, but results from them indicate that this might not lead to the architectural disaster that one would expect. Coordinated algorithms, it transpires, are very resilient to the addition of random behavioural rules. The reason for this is that rogue builders can rarely find a configuration in the virtual nest to which the random rule is applicable, so the coherence of the architecture remains. This has interesting implications for the evolution of differences of nest architecture between wasp species. As related coordinated algorithms produce similar yet distinctive architecture, it seems possible that individuals in colonies producing a particular nest design could accumulate additional building rules that had no visible effect on the completed nest. A new nest architecture might then appear in a single generation when the recombination of genes in offspring of parents from two such colonies creates a new coordinated algorithm.

So, the pattern of evolution of social insect nests may well, as Richard Dawkins suggested back in 1982, be different for changes in material technology compared to design. It is surprising that these important speculations, and indeed others relating to animal building behaviour, arising from *The Extended Phenotype* have not yet stimulated the research effort they deserve. It is a book that particularly speaks to researchers; there is valuable detail in its message. On the other hand there are examples where we can clearly see that Dawkins' way of explaining the evolutionary process has led to new understanding. It is on such an example that I will end, throwing in some research data by Charles Darwin to set the scene.

Darwin, among his varied output, published a treatise on earthworms; this includes the calculation that, through their little piles of excreta or casts, they can bring to the surface at least 8.4 pounds of soil per square yard per year. Two modern preoccupations have grown out of this and related observations: *ecosystem engineering*, which is concerned with the ecological implications of the ways organisms modify their habitats, and *niche construction*, which is particularly concerned with evolutionary consequences such as modification. Here, it is the latter that interests us.

I felt comfortable in dismissing the footprints of the Carboniferous 'centipede' as an extended phenotype, and might dismiss a worm cast for the same reason; how can a worm cast phenotype influence the survival of worm-cast-making genes in the burrowing worm? Darwin's calculations force me to think again. The cumulative effect of casts created by a population of earthworms clearly does change the worm environment in a significant way, and indeed that of other organisms in the same habitat, most obviously of plants. Not only that, this environmental change persists, to be inherited by their earthworm offspring. So, worms inherit genes for making burrows, but they also inherit a world altered by previous burrowing genes. What are the evolutionary implications of this?

This is just the sort of problem that mathematical modellers can help us to understand and the first steps in modelling the consequences of environmental inheritance have been tested, in 2003, by Odling-Smee, Laland, and Feldman. Their model imagines a simplified world in which to study the replication of alleles at only two gene locations. One of these locations, E, carries alleles that bring about changes in the environment. These affect the amount of some resource R in the environment that in turn determines the replication success of alleles at a second gene location, let's call it A. The resource R could, as its name seems to imply, be food. So, there might be two possible alleles at E, one altering the environment to enhance food availability, while the other does not. But the same argument would apply if the 'resource' was predators, where one allele for building behaviour genes changes the environment in a way that increases protection from predators while the other does not.

Let's imagine, for example, that the environmental change is very small, building up as each succeeding generation makes its contribution. The model shows, perhaps not surprisingly, that the greater the number of previous generations that are necessary to alter R through environmental inheritance, the greater the time lag in the effect on A; but there are less obvious findings too. When E starts to exert its effect on R, the delay may be so great that, although an allele at A will ultimately benefit, its frequency initially continues to decline and may be lost from the population altogether before the environment is sufficiently changed so that selection favouring it takes effect.

Don't worry if this is beginning to sound rather technical, just look at the language. It is the language of *the gene as the unit of selection*. It is also the language of *the long reach of the gene*, since this model can make predictions on the relative success of alleles at locus A in one type of organism (grass, for example) due to the influence on the environment brought about by genes at the E location in quite different organisms (earthworms). *The Extended Phenotype* following on from *The Selfish Gene* was a landmark in persuading biologists and the general public to talk and think in this way. I am also personally indebted to Richard, who has helped my ideas and understanding to travel a long way since asking 'What is a puma?'

THE SELFISH GENE

LIVING WITH
THE SELFISH GENE

Marian Stamp Dawkins

I HAVEN'T lived with the author of *The Selfish Gene* for nearly twenty-five years but I have lived in close and constant contact with the book itself since before it was published. Its words and figures of speech are thrown at me almost daily from student essays. The questions it raises are the driving force behind the most successful tutorials and the sound of pennies dropping as each new generation of students takes in the extraordinary implications of what the book is saying, is as loud now as it ever was. *The Selfish Gene* seems to occupy a unique place in biological writing. Some books, like Rachel Carson's *Silent Spring*, herald a new age and have a huge impact on the way people think at the time but thereafter are read mainly for historical interest. Others form part of an ongoing movement but are soon superceded by more up-to-date versions or more fashionable means of expression. But if *The Selfish Gene* had not been written when it was, there would still be a need for it to be written today. There are simply no books that have taken its place, even now when so many other books have followed in its wake.

What follows are a few musings on *The Selfish Gene* as the most important teaching aid I have ever come across. It has the power of educating in the original meaning of the word—it literally leads people out of one way of thinking and forces them, often quite painfully and agonizingly, to see the world in a different way. It confuses, it disturbs, it offends, even. It makes people see that what they may have believed before is not compatible with a gene-centred view of the world, so that they either have to rearrange their minds to accommodate a new reality or to think more logically about what they do believe as a viable alternative. Some people love it. Some people are shocked. But no one who reads it and fully takes in what it is saying is ever quite the same again.

The first advantage of teaching with *The Selfish Gene* is that it instantly solves the problem of detecting plagiarism in student essays. Students often take short cuts and copy out paragraphs of papers or textbooks as part of their essays, imagining that this will not be noticed. Until there was software to help you locate the source, you had to rely on your own memory of exactly what different books had said to catch them at it. But they could never get away with it if they tried it with *The Selfish Gene*. Laboured student prose that suddenly bursts forth with words such as 'vehicles' and 'immortal coils' immediately gives itself away. I used to try and deal with such 'borrowings' by saying (with a singular lack of originality on my part), 'I see you agree with Richard Dawkins' but gave that up because I never once met a student who was in the least embarrassed or contrite. (I suppose the ones who would realize I was trying to be ironic would not have been quite so blatant in the paragraphs they copied out.) So I then switched tactics and discovered an even more powerful way in which *The Selfish Gene* educates the unwary.

The language of *The Selfish Gene* so seductively captures exciting new concepts in a few well chosen words that at first people don't realize that they are being enticed into a gentle trap of their own making. Captivated by the poetic way in which difficult ideas are described, they swallow them whole and later regurgitate them, undigested, into their own speech and writing. That is when the trap closes and the real education begins. Simple questions such as 'What exactly do you mean by describing an elephant as a vehicle for genes?' can be devastating and far more difficult to answer than 'What do you think Dawkins means by . . .?' because the ideas are no longer those of some distant author who may have gone too far. By using the ideas themselves, people have owned them. They have skated out onto thin ice of their own accord and then find there is no escape from the difficult process of finding their own way on to firm ground. That is real education—beginning to think for yourself and finding the intellectual courage to negotiate both the doubts and the certainties that result.

The Selfish Gene makes it possible to be a Devil's Advocate of a most gentle kind, using doubt and confusion as constructive educational tools rather than as destructive ones. How can you possibly say that genes—as bits of DNA—are selfish? Are you really saying that parents look after their young because that's the way genes for parental care spread themselves? Come to that, how on earth can you say that there are genes for behaviour at all when we know no genes act in isolation? The great thing about putting such questions to someone who claims to have read and agreed with *The Selfish Gene* is that the answers to most of them are there, in the book itself. By asking awkward questions, you are not casting your students adrift in a sea of confusion with no maps. You have given them a guidebook which, despite all the criticisms that have been heaped on it, actually contains the answers to most of its critics and certainly to most of the questions that are usually raised about it. Add the first two chapters of *The Extended Phenotype* and you have a comprehensive guide to what a gene-centred view of evolution is and is not about. I know of no other book that not only makes people think and rethink but cushions the uncomfortable process of their own doubting quite as comprehensively and constructively.

(The mention of guidebooks and journeys has incidentally reminded me of a less serious memory of living with *The Selfish Gene*. The Zoology Department in Oxford used to have a tradition of having comic sketches at its Christmas party in which graduate students would impersonate various senior members of staff. One year there was a sketch in which Richard was depicted clutching a large illustrated edition of his book, waving his arms in the air and insisting on taking everybody on journeys all over the place. I can't remember whether he thought it was as funny as the rest of us did. But I digress.)

A related and more worrying feature of life with *The Selfish*

Gene that never fails to astonish me, and constantly reinforces the point that it is a book for the present and the future not just the past, is that many of the lessons that I thought would have been learned a long time ago still haven't been, even among professional biologists. We still need *The Selfish Gene* to explain the widespread confusion that surrounds controversies such as the level at which selection acts, the interaction between genes and culture, the fact that natural selection doesn't necessarily lead to perfection and so on. But above all we need it to explain, yet again, that believing in the power of natural selection does not imply that genes determine everything we do. Adaptation by natural selection implies that there is or was underlying heritable variation for natural selection to get its teeth into, but genes have many and various ways of using the phenotype to have their effects (as explained more fully in that much underrated book, *The Extended Phenotype*). Plants with a gene for extra growth (perhaps an extra shot of growth hormone), for example, will not necessarily grow all that tall if they are planted in poor soil. They might even turn out to be smaller than plants without this 'tallness' gene that are grown in rich soil. Genes are passed into the next generation by an incredibly complex interaction with other genes and by making the body they are in interact with, and take in information from, its environment. The old idea that 'genetic' meant 'fixed' or 'inevitable' has long ago been replaced by a more realistic idea of the importance of flexibility. Or so I keep hoping.

Only recently, however, an undergraduate confidently told me that Richard was a genetic determinist. She knew this because she had been to a seminar in London which said so. She had not read *The Selfish Gene*. Upon further enquiry, it turned out that the seminar had been given by a prominent biologist who, if her account were to be believed, had not read *The Selfish Gene* either. I suggested that she went away and read it (and read Chapter 2 of *the Extended Phenotype*). Our next conversation was quite different. She hadn't realized how people had distorted Richard's views, and I hadn't either. It made me realize how important it is that people keep reading *The Selfish Gene*, not just reading it once or through the criticisms of other people but honestly taking the trouble to understand the revolutionary view of the world that it portrayed in 1976 and still gives us, with a unique vision of undiminished freshness, thirty years later.

* * *

In the summer of 2005, an Oxford student sat down to write a final examination paper. The first question they chose to answer was about why some animals do not reproduce and how natural selection could possibly favour them if they did not. The student wrote a scholarly essay, citing many different authors and relevant examples. Technical terms were correctly defined and used but there was something familiar about the style. With unusual frankness, he/she (the exam papers were of course only known by numbers) remarked at one point in the essay: 'And here I rely heavily on the words of Richard Dawkins'. Yes, I thought as I marked the script and gave the candidate a high mark not just for the content of the essay but for actually acknowledging their source. Yes. Don't we all?

THE GENE MEME

David Haig

IN the final chapter of *The Selfish Gene*, Richard Dawkins explored the analogy between genetic and cultural evolution. Cultural traits, he suggested, evolve by a process of natural selection in which there is preferential proliferation of traits with properties that promote their own transmission. 'We need a name for the new replicator,' he wrote, 'a noun which conveys the idea of a unit of cultural transmission, or a unit of imitation. "*Mimeme*" comes from a suitable Greek root, but I want a mono-syllable that sounds a bit like "gene". I hope my classicist friends will forgive me if I abbreviate mimeme to *meme*.' Dawkins concluded his discussion, 'However speculative my development of the theory of memes may be, there is one serious point which I would like to emphasize once again. This is that when we look at the evolution of cultural traits and at their survival value, we must be clear *whose* survival we are talking about.' He entertained the possibility 'that a cultural trait may have evolved in the way that it has, simply because it is *advantageous to itself*'.[1]

The 'meme' has exhibited admirable powers of replication and persistence in the thirty years since its conception, but its cultural spread pales before that of the monosyllable it was chosen to imitate. In the first half of this essay, I will consider the diverse meanings that have become associated with that simple meme, the 'gene'. Not every scientist means the same thing when they refer to a gene and these differences in nuance can be a source of confusion. In particular, I will discuss Dawkins' explicit definitions of the selfish gene and, in the guise of the *strategic gene*, propose what I believe to have been Dawkins' implicit definition of the selfish gene. We can think of the changing and diversifying concepts of the gene as an example of memetic evolution. The second half of this essay will use the discussion of the 'gene' in the first half to illuminate the status of the 'meme' as a putative replicator subject to natural selection.

The word 'gene' was introduced into the English language by the Danish plant breeder Wilhelm Johannsen in an address before the American Society of Naturalists in December 1910. His intent was to take issue with the common 'conception that the personal qualities of any individual organism are the true heritable elements or traits!'[2] The rediscovery of Mendelism had shown that 'the *personal qualities* of any individual organism do not at all cause the qualities of its offspring; but the qualities of both ancestor and descendant are in quite the same manner determined by the nature of the "sexual substances"—i.e., the gametes—from which they have developed. Personal qualities are then *the reactions of the gametes* joining to form a zygote; but the nature of the gametes is not determined by the personal qualities of the parents or ancestors in question.' Thus, Johannsen made a crucial distinction between *phenotype* (observable traits) and *genotype* (heritable factors).

In Johannsen's view, the mistaken notion of the inheritance of personal qualities was reinforced by the persistence of an outdated vocabulary. 'It is a well-established fact that language is not only our servant, when we wish to express—or even conceal—our thoughts, but that it may also be our master, overpowering us by means of the notions attached to the current words. This fact is the reason why it is desirable to create a new terminology in all cases where new or revised conceptions are being developed. . . . Therefore I have proposed the terms "gene" and "genotype" and some further terms, as "phenotype" and "biotype", to be used in the science of genetics. The "gene" is nothing but a very applicable little word, easily combined with others, and hence it may be useful as an expression for the "unit-factors", "elements" or "allelomorphs" in the gametes, demonstrated by modern Mendelian researches.'

From this beginning, Johannsen's gene has had an illustrious history (as have 'genotype' and 'phenotype', but not 'biotype').

But 'gene' itself conveys little information, consisting as it does of only four letters and a single syllable when spoken. The factors accounting for its success as a meme are probably those identified by Johannsen—that it was 'a very applicable little word, easily combined with others'—and the historical contingency that the word was used to represent a set of ideas and concepts that had high memetic fitness. If 'gene' is a meme, it is a rather uninteresting one. The interesting memes are the shifting concepts of the units of inheritance for which gene was a convenient label. The memetic history of 'gene' is interesting only in so far as it provides a marker for the propagation of these more amorphous ideas and concepts as they have undergone constant reformulation.

'Gene' has never had a single meaning, but has always had different meanings to different people, and often different meanings for a single person, depending on context. For each person who added gene to their vocabulary, the word had a meaning that was derived from explicit definitions either read or heard, inferences from how the word was used, and reformulations of the concept within their own minds. This private definition of gene was then translated into new definitions and new uses in conversation and writing that were perceived by other minds and incorporated into new private definitions. My intention in stating the obvious is to point out what must be true of most memetic transmission: there is some degree of continuity in the propagation of ideas from mind to mind, but it lacks the high fidelity of the propagation of genes from generation to generation.

Johannsen, of course, had his own conception that he wished to convey to others. 'As to the nature of the "genes" it is as yet of no value to propose any hypothesis; but that the notion "gene" covers a reality is evident from Mendelism. . . . We do not know a "geno-type", but we are able to demonstrate "genotypical" differences or accordances. . . . genotypes can be examined only by the qualities and reactions of the organisms in question.' Genes are known by their phenotypic effects. Johannsen was dismissive of attempts to localize genes. 'The question of *chromosomes* as the presumed "bearers of hereditary qualities" seems to be an idle one. I am not able to see any reason for localizing "the factors of heredity" (i.e., the genotypical constitution) in the nuclei. The organism is in its totality penetrated and stamped by its genotype-constitution. All living parts of the individual are potentially equivalent as to genotype-constitution.'

Johannsen's 'applicable little word' soon gained wide currency among geneticists, especially among those who believed that the gene corresponded to a physical structure on chromosomes. (One could say there had been memetic recombination that attached gene, as a label, to an alternative concept of the unit of inheritance.) Supporters of the chromosomal theory, however, continued to define the gene operationally as that which was responsible for a heritable phenotypic difference. A. H. Sturtevant, one of the first to map genes to chromosomes, commented in 1915: 'We can . . . in no sense identify a given gene with the red colour of the eye, even though there is a single gene differentiating it from the colourless eye. So it is for all characters . . . All that we mean when we speak of a gene for pink eyes is, a gene which differentiates a pink eyed fly from a normal one—not a gene which produces pink eyes per se, for the character pink eyes is dependent on the action of many other genes'.[3]

Much of twentieth-century experimental genetics was engaged in making inferences about the physical nature of genes from observations of their phenotypic effects (i.e., by observations of differences in an organism's physical characteristics). These studies led to a definition of the gene as a stretch of DNA sequence that is responsible for specifying the amino acid sequence of a protein. Thus, the operational definition of a gene—that a gene is known by its effects—began to shift as genes came to be viewed as tangible elements with defined chemical properties. Now, the existence of a gene is often inferred from properties of a DNA sequence without any information about the gene's phenotypic effects and without the observation of differences among sequences. But the definition of the gene as a protein-encoding stretch of DNA is more recent than its definition as that which is responsible for a phenotypic difference, and it is not surprising that the modern molecular definition has not fully supplanted the older operational definition.

Experimental geneticists invoke genes to explain *observed* phenotypic differences. A pink-eyed fly differs from a red-eyed fly because the former possesses a gene for pink eyes inherited from both parents whereas the latter has inherited at least one gene for red eyes. In a similar way, evolutionary biologists often invoke genes to explain *hypothetical* phenotypic differences in an attempt to understand the nature of adaptation by natural selection. An ornithologist might wish to understand why males of some species help to raise offspring (dads) while males of other species put all their efforts into seeking additional copulations (cads). She might posit a gene for being a cad and ask under what circumstances it would invade a population in which most males behave as dads. To paraphrase Sturtevant, all that we mean when we speak of a gene *for* being a cad is a gene which differentiates a cad from a dad—not a gene which produces caddish behaviour per se, for caddishness results from the action of many genes.

The use of genes to explain both differences among individuals and invariant features of organisms has led to an unfortunate confusion in public discourse over claims that genes cause behaviour. Consider a contentious example. Behavioural geneticists are interested in differences among individuals and look for genetic factors that might explain why some people, but not others, engage in violent acts. Perhaps violent offenders are unable to control their impulses because they carry a mutation in the gene encoding the enzyme monoamine oxidase. The explicit comparison is between the behaviours of individuals with different variants of the gene. Evolutionary psychologists, on the other hand, are interested in species-typical behaviours that they view as adaptations to enhance the survival and reproduction (reproductive fitness) of individuals. Therefore, they seek explanations for why we have evolved a genotype that makes us more likely to engage in violence under some circumstances, but not others. Perhaps young men are predisposed to violent behaviour when they control few resources in societies in which there is a large gap between rich and poor. The implicit comparison is between the reproductive fitness of individuals in the present world (or in a past environment in which the behaviours were adaptive) and the reproductive fitness of individuals in alternative worlds in which genes respond differently to the environment. Thus, behavioural geneticists ascribe the observed difference between offenders and non-offenders to a genetic difference, whereas evolutionary psychologists ascribe the same difference to environmental factors. Yet both groups are castigated as 'genetic determinists' by those who reject biological explanations of human behaviour.

How then did Dawkins define the eponymous protagonist of *The Selfish Gene*? Dawkins recognized that there is 'no universally agreed definition of a gene. Even if there were, there is nothing sacred about definitions. We can define a word how we like for our own purposes provided we do so clearly and unambiguously. The definition I want to use comes from G. C. Williams. A gene is defined as any portion of chromosomal material that potentially lasts for enough generations to serve as a unit of natural selection'.[4] Thus, the gene could be a longer or shorter unit than the protein-encoding gene recognized by molecular biologists. When defined in this way, Dawkins believed that the gene must be recognized as 'the fundamental unit of natural selection, and therefore the fundamental unit of self-interest'.

Dawkins' recognition of the gene as the fundamental unit of selection has not met with universal acceptance, however, partly because different scientists have different implicit definitions of the gene. David Sloan Wilson, for example, has argued that the gene deserves no special status because it is merely the lowest level of a nested hierarchy of units of selection: genes, cells, individuals, groups, species. In this view, genes are nested within cells, cells within individuals, and individuals within groups; and natural selection can act at all levels of the hierarchy. Thus, there may be adaptations for the good of individuals and groups as well as of genes. The placement of the gene at a level of the hierarchy below the cell implicitly defines the gene as a material object located within cells.[5] But this is not Dawkins' concept: 'What is the selfish gene? It is not just one single physical bit of DNA . . . it is *all replicas* of a particular bit of DNA distributed throughout the world. . . . The key point . . . is that a gene might be able to assist replicas of itself which are sitting in other bodies. If so, this would appear as individual altruism but it would be brought about by gene selfishness.' For Wilson, the gene is a *material* object that resides within cells whereas for Dawkins the gene is a piece of *information* distributed across multiple levels of Wilson's hierarchy. (I cannot resist suggesting the usit—pronounced 'use it'—as an applicable little term to represent the disputed unit-of-selection whatever that may be and mean.)

The continued debate between gene-selectionists and group-selectionists identifies an ambiguity in the meaning of the word 'gene', even when this is defined as a rarely-recombining stretch of DNA. A gene could refer to the group of atoms that is organized into a particular DNA sequence—each time the double helix replicates, the gene is replaced by two new genes—or it could refer to the abstract sequence that remains the same gene no matter how many times the sequence is replicated. We might call these concepts the *material gene* and the *informational gene*. Dawkins refers to something like the informational gene when he describes the selfish gene as '*all replicas* of a particular bit of DNA' but I believe that he neither wanted nor intended this definition. If all humans came to share the same DNA sequence, the theory of the selfish gene would not predict universal benevolence. A selfish gene does not 'care' about all replicas of its sequence, but only about some of its replicas in a smaller group of related individuals. The reason why not is tied up with the dynamics of genetic replicators.

Every genetic novelty (new informational gene) originates as a modification of an existing informational gene and is initially restricted to a relatively small number of material genes. Therefore, the informational gene's material copies will interact with each other only when they are present in different cells of the same body or in the bodies of closely-related individuals. If such a gene is ever to become established, it must be able to increase in frequency under these circumstances. As the gene's frequency increases, its fate may be influenced by selection at higher levels of the material hierarchy, but it will still retain the features that ensured its success when rare. Thus, the gene can be said to commit itself to a strategy when rare that it must maintain at all frequencies. The phenotypic effects of successful genes will consequently appear to be adaptations for the good of groups of material genes that interact because of recent common descent. It is this coterie of material genes that is the *unit of adaptive innovation.* I will refer to such a group of adaptively-interacting material genes as the *strategic gene* because it is this unit that is the strategist in an evolutionary game played against similar groups of material genes.[6]

A material gene has dual roles. It can be *expressed*—that is, its sequence can be transcribed into a messenger RNA that is translated into a protein—and it can be *replicated* to produce copies of itself. The essence of adaptation by natural selection is that a material gene's phenotypic effects, such as the protein it encodes, influence the probability that the material gene, or its replicas, will be copied. The extent of the strategic gene is determined by the number of replication cycles that separate the material genes responsible for the expression of a phenotypic effect from the material genes that thereby have an increased probability of being copied. Thus, the strategic gene is not a fixed entity but can evolve to encompass more, or fewer, material copies of an informational gene.

Take, for example, a large, well-mixed population of single-celled phytoplankton. Once a cell divides, its two daughter cells separate and never interact again, except by chance. Each material gene is subject to selection solely on how its own expression influences its own replication. In this case, the strategic gene is limited to a single material gene. Now consider a cod fish in which multiple individuals of both sexes spawn (release eggs or sperm) simultaneously. A sperm and egg fuse to form a zygote that develops into a large multicellular individual that may itself contribute eggs or sperm to zygotes of the next generation. Zygotes become widely dispersed by ocean currents so that there are no preferential associations among kin. A single material gene in a zygote gives rise to replicas in all the cells of an adult cod. Material genes in the heart and brain of the cod never replicate, yet their expression promotes the replication of their replicas in the cod's gonads. In this example, the strategic gene is spread throughout the body of a single fish. Finally, consider a beehive. Material genes that are expressed in the hive's sterile workers promote the replication of their copies in the ovary or sperm storage organs of the queen bee. So, in our third example, the material copies of the strategic gene are spread among the members of the hive. When Dawkins discusses selfish genes, it is in the sense of the strategic gene described above that the concept should be understood.

Dawkins proposed that memes play an analogous role in cultural evolution to that played by genes in biological evolution. If so, memes should display features that promote their own replication. Such features could be interpreted as adaptations 'for the good of' the meme itself. In the remainder of this essay I shall look at the analogy between genes and memes. I will employ a vague definition of a meme as 'a mental item that is borrowed from one person and passed on to another'. There are many things that could be considered to be memes, but my focus will be on the transmission of ideas, and I will use the concept of 'the gene' as an exemplar.

Rather than looking directly at who benefits from memetic transmission, let's consider who benefits from communication. Many communication acts are committed because a sender wants to produce some change in a receiver. Such acts can be considered to be propaganda, from the Latin for propagation. An item of propaganda, a *propagandum*, is a device designed by a propagandist to achieve a change in the actions of a receiver. The propagandum has served *the purpose of the propagandist* if the receiver acts in the desired manner. To achieve this purpose it isn't necessary that a receiver pass the propagandum on to others. But if there is no chain of transmission, then the propagandum does not qualify as a meme. Its effects are not advantageous to itself, but only to the propagandist.

Sometimes a propagandum is designed to be passed from one receiver to another because this increases the propagandum's efficiency as an agent of mass persuasion. If the propagandist is successful in achieving ongoing transmission, the propagandum then qualifies as a meme. The propagandum serves its designer's ends if receivers act in some desired manner and if receivers pass the propagandum on to others to affect their behaviour. The features that promote the propagandum's transmission benefit the propagandist but can also be said to benefit the propagandum, considered as a meme. But the features that effect a change in the behaviour of receivers, while they benefit the propagandist, need not benefit the meme.

Of course a propagandist's designs may misfire. A propagandum could fail to achieve the propagandist's ultimate purpose of changing behaviour, but could succeed in the subsidiary purpose of being propagated from mind to mind; or, the propagandum could continue to propagate from mind to mind after it no longer serves the propagandist's goals. Once a chain of memetic transmission exists, there is no selection on a propagandum to serve its original designer's ends, although the propagandum, as meme, may continue to serve these ends if the fidelity of transmission is sufficiently high.

Each step in a chain of transmission is an act of *selection*, in so far as the transmitter chooses to transmit one meme rather than another (or no meme at all). We can think of any feature of a meme that has predisposed successive transmitters to 'want' to pass it on as an adaptation of the meme to enhance its own transmission: such adaptations can appeal to either the conscious motivations or the unconscious motivations and biases of transmitters; and such adaptations can be intended features, consciously selected by a propagandist, or they might be unintended features that arise from the interaction of 'random' mutation with differential replication during the chain of transmission. That is, the adaptive features of memes may be the products of 'intelligent design', 'natural selection', or a combination of the two.

Whose interests, then, are served when a meme is transmitted? We can look at this from two perspectives: that of individuals, and that of memes. First, we need to consider the interests of the transmitters at each step in the chain. If a person consciously chooses to transmit a meme, the meme must serve some perceived interest of theirs. The choice serves a *perceived* interest because individuals can be mistaken about what will promote their true interests. For example, a meme may be a propagandum serving the *actual* interests of some other person earlier in the chain. (By the interests of a person, I mean here their own self-defined goals in life.) Second, there are the metaphorical interests of the meme in its own transmission.

Does taking the second perspective, of viewing culture through the lens of a meme's interests, contribute anything that could not be obtained from the first perspective? A meme's-eye view could be justified if it were shown that there are features of memes that promote a meme's own interests without serving the interests of any of the meme's transmitters. Such features might appeal to quirks of nervous systems that are better considered as unconscious biases rather than sources of personal motivation. A preference for a meme's-eye view might also be defended if the features that make a meme likely to be passed on had accumulated in many steps over the course of memetic transmission.

Johanssen invented the 'gene' to clarify the distinction between genotype (gene) and phenotype (trait). Can a similar distinction be made for a science of memetics? There are two principal kinds of things we observe that provide evidence about the nature of memetic transmission. The first are communication acts including sounds, texts, actions, and artefacts. The second are insights from introspection when we register a communication act, when we integrate the content of a communication act into our private set of concepts, and when we emit communication acts. Introspection may be an unreliable guide because unconscious aspects of our motivations are hidden and our conscious perceptions may be partial, inaccurate, and misleading. Communication acts appear closer to the concept of genotype (things transmitted) whereas the conscious and unconscious effects of these acts on our internal state appear closer to phenotype (effects that influence what is transmitted). In the history of genetics, the phenotype was apparent and the genotype hidden. But this relation seems to be reversed for memetics. Memes are observed, rather than inferred from their effects, whereas their effects are in large part hidden.

The phenotype/genotype distinction works fairly well for genes, but there are many unresolved problems in its application to memes. For example, let's suppose that there is a body of lore preserved and updated by the Medieval Guild of Propagandists about what techniques are effective in changing public opinion, and that this body of lore is passed from master to apprentice. The apprentices use proven techniques to design propaganda and then a propagandum's success in public persuasion influences whether the apprentice passes on the technique to his own apprentices when he becomes a master. From the perspective of the techniques, considered as memes, propaganda items are meme-products that influence a technique's probability of transmission, but these items may also function as memes in their own right. A propagandum may be both 'memotype' and 'phemotype'.

The gene has a material definition in terms of a DNA sequence that maintains an uninterrupted physical integrity in its transmission from generation to generation. Memes also have a physical form in their transmission from one individual to another, sometimes as sound vibrations, or text on paper, or electronic signals relayed through a modem. When these 'outward' forms of a meme are perceived, they elicit changes in a nervous system that constitute the meme's 'cryptic' form. The material basis of the cryptic form is probably unique to each nervous system colonized by the meme. Memetic replication, then, has nothing like the elegant simplicity of the double helix.

If the material form of memes is problematic, might it be more appropriate to define memes purely in terms of information? But what are the memes in our evolving concepts of the gene? These

concepts have been reformulated and recombined with other ideas
at each step in the chain of transmission. How can one identify the
'nuggets' of ideas that remain unchanged during this process and
thus persist 'for enough generations to serve as a unit of natural
selection'?

Dawkins argued that 'selfishness is to be expected in any entity
which deserves the title of a basic unit of natural selection'. His
definition of the gene qualified as such an entity because it pos-
sessed three properties that 'a successful unit of natural selection
must have . . . longevity, fecundity, and copying fidelity'. Unlike
his careful definition of the gene, Dawkins was somewhat vague
about the definition of a meme, simply stating that this was a 'unit
of cultural transmission, or unit of imitation'. Is there some way
to define a meme so that it possesses the properties that would
qualify it as a unit of natural selection (and hence deserving of the
'selfish' label)?

Consider the 215 pages of the first edition of *The Selfish Gene*.
Dawkins' slim volume contains many ideas influenced by older
texts and itself has influenced ideas expressed in newer texts
(including this one). Can *The Selfish Gene* be parsed into a set of
selfish memes each displaying longevity, fecundity, and copying
fidelity? Or do we need to parse the text? Dawkins' definition of
the gene did not specify boundaries between genes. The gene was
a piece of chromosome sufficiently short to last long enough,
without recombination, to function as a unit of selection. But this
definition meant that there were many different, overlapping ways
in which a chromosome could be divided into genes. Could the
same approach work for memes?

Dawkins' principal interest was in the phenotypes of organisms
rather than of genes. Just as he failed to specify precisely how to
divide a chromosome into genes, he did not specify how to divide
the phenotype up into individual adaptations due to individual
genes. I believe this approach was justified for his purposes. As
long as all parts of the genome have the same rules of inheritance,
what is good for one part of the genome is also good for any other
part of the genome, and the genome itself can be considered as an

adaptive unit. Highly complex adaptations require a long genetic text. There are two widespread solutions to this problem that one can call the asexual and the sexual solution. In the asexual solution, an entire genome replicates at the same time and does not recombine with other genomes. Thus, the whole genome behaves as a single Dawkinsian gene. In the sexual solution, two entire genomes come together for some length of time, then separate into two new genomes after exchanging interchangeable parts with each new genome receiving one of each part. The genome is an ephemeral collective of many Dawkinsian genes, but the rules of Mendelian inheritance ensure that what is good for one part is good for all, at least for the time that the genes are temporarily associated. (I leave to one side the complexities that arise when the 'rules' are broken and there is conflict within the genome.)

Neither the sexual nor the asexual solution seems to apply to most complex memetic 'texts'. Ideas recombine freely to generate each new text and there is no well-defined exchange of interchangeable parts. One idea can be adopted from a text and the remainder abandoned. Therefore, the adaptations of memes will be adaptations for the good of the individual, rarely-recombining ideas. Some of these ideas may be so simple—the idea that the gene is a part of a chromosome—that they can exhibit few, if any, adaptations for their own transmission. I see little value in treating such ideas as selfish, just as there is little value in treating a single nucleotide as selfish. Such ideas serve the utility of propagandists or (perhaps) of larger non-recombining meme complexes into which they become incorporated. The place to look for sophisticated adaptation and selfishness will be in coherent ideologies, large 'asexual' meme complexes that are transmitted as a unit with high fidelity of transmission. Richard Dawkins would identify the world's great religions as the prime examples and he would argue that the free recombination of ideas is important if ideas are to serve our ends rather than their own.

These are some of the problems I see in defining memes and thinking of memes as selfish. And yet, in the twenty-five or more years since I first read *The Selfish Gene*, there is no section of the book that has stayed more in my mind than its final chapter. In the intervening period, I have passed on the selfish meme many times in conversation, and here I am writing an essay on memes. The meme of the 'meme' is a tenacious beast, at least for those minds that are vulnerable to its charms. The current essay is a work of propaganda. I wish to communicate ideas that I hope will influence your own concepts of genes and memes. If I am effective, you may pass on these ideas in modified form to others. In pursuit of these ends, I have crafted phrases to *grab your attention*, and have worked, and reworked, on clarifying concepts in my own mind. This process has involved testing numerous alternatives against the standards of what I think will be effective and what will form a coherent whole with the rest of the essay. I think of the essay as part of my own intention and not of the ideas that it tries to communicate. But am I fully autonomous in this process? Many ideas have competed for inclusion during the course of writing, but only some have made it into a final version that has nothing like the form and content that I intended when I first sat down to write. It is only in retrospect that I know what I have chosen to write. The final version contains the ideas that have *grabbed my attention*. It has sometimes seemed that they are using me for their ends. What fraction of these ideas are my own and what fraction have been borrowed from others? The web of intellectual influence is complex and it is unclear whether I ever have a truly original idea.

The Oxford biologist-cum-psychologist George Romanes wrote in *Darwin, and After Darwin*:

Quite apart from any question as to the hereditary transmission of acquired characters, we have in this *intellectual* transmission of acquired *experience* a means of accumulative cultivation quite beyond our powers to estimate. For, . . . in this case the effects of special cultivation do not end with the individual life, but are carried on and on through successive generations *ad infinitum*. . . . [In] this unique department of purely intellectual transmission, a kind of non-physical natural selection is perpetually engaged in producing the best results. For here a struggle for existence is constantly taking place among 'ideas', 'methods', and so forth, in what may be termed a psychological environment. The less fit are superseded by the more fit, and this not only in the mind of the individual, but through language and literature, still more in the mind of the race.[7]

Richard Dawkins exhorts us to ask: 'fit' in what sense and for whom?

ENDNOTES

1 References to Richard Dawkins' writings in this chapter are taken from R. Dawkins, *The Selfish Gene* (Oxford: Oxford University Press, 1st edn., 1976).

2 W. Johannsen, 'The genotype conception of heredity', *American Naturalist*, 45 (1911): 129–159.

3 A. H. Sturtevant, 'The behavior of the chromosomes as studied through linkage', *Zeitschrift für induktive Abstammungs- und Vererbungslehre*, 13 (1915): 234–287.

4 The source of Dawkins' definition is G. C. Williams, *Adaptation and natural selection* (Princeton: Princeton University Press, 1966).

5 D. S. Wilson and E. Sober, 'Reintroducing group selection to the human behavioral sciences', *Behavioral and Brain Sciences*, 17 (1994): 585–654.

6 D. Haig, 'The Social Gene', in J. R. Krebs and N. B. Davies (eds.), *Behavioural Ecology: An Evolutionary Approach* (Oxford: Blackwell, 4th edn., 1997), 284–304.

7 G. J. Romanes, *Darwin, and After Darwin. II. Post-Darwinian questions: Heredity and Utility* (Chicago: Open Court, 1895), vol. 2, page 32.

THE INTELLECTUAL CONTRIBUTION OF *THE SELFISH GENE* TO EVOLUTIONARY THEORY

Alan Grafen

A PHENOMENON such as Dawkins' *The Selfish Gene*[1] can be seen from many points of view and set in many contexts. Its popular success, its influence on generations of students and scholars, and its permeation of the intellectual life of many countries could all be taken as starting points. Instead, this article will begin by focusing very narrowly on the originality and intellectual significance of the ideas in *The Selfish Gene*, and how they stand up today in the light of further research. I will consider the book's reception, and give reasons for the variability of esteem in which it is held.

For my purposes, the core arguments of *The Selfish Gene* are (i) the introduction of the concept of a replicator, which allows what was then the most logically rigorous exposition of Darwin's theory of natural selection, (ii) the link between replicator selection and selfishness, in a technical sense, and (iii) the suite of links that establish in turn each of the then new theoretical ideas in adaptationist biology in terms of the selection of replicators.

George Williams[2] had developed Darwin's argument of natural selection by verbal and conceptual means, but Dawkins' forensic analysis in terms of replicators provided for biology a new understanding of the Darwinian logic. The properties of fidelity, fecundity, and longevity explained what kinds of objects could be replicators, and why DNA was such a powerful one. This was new theory in biology, and it was followed by another crucial breakthrough. The term 'selfish' has been the subject of much debate, but by bringing it so centrally into play, and by explaining so clearly and carefully what it meant, Dawkins was taking a formal step in the understanding of the nature of adaptation. The idea of adaptation was an old one, predating Darwin. Again building on Williams' ideas, Dawkins pointed out, though not in these terms, that logically the concept of selfishness makes sense only within a larger set of ideas, in which there is an entity to be selfish, a quantity whose numerical value says how well-off the entity is, and a range of possible actions that can be taken by the entity. Mathematics and the social sciences nowadays use mathematical formalizations of this larger context ('optimization programmes') in game theory and economics, to cope with familiar economic issues such as consumer choice theory and profit maximization by firms. By bringing this larger set of ideas into the centre of the biological argument, Dawkins was able to pursue the logic of adaptationism further. Linking selfishness to replicators, he argued that the entity concerned was the gene and the quantity that surviving genes would come through natural selection to act as if maximizing was their replication. This link between replicator dynamics on the one hand and selfishness on the other, brilliantly encompassed in the title of the book, is the very centre of Darwin's argument, but spelled out in a way that is more practical for further analysis: of course, it could only be articulated after Mendelian genetics had been discovered.

But Dawkins did not stop there. His triumph was to take the various recent flowerings of adaptationism, and to establish their unity under Darwinism by interpreting them all in terms of the logical framework of replicators. First and foremost, W. D. Hamilton's inclusive fitness[3] worked through a replicator assisting copies of itself. Characteristically Hamilton had himself already explained his ideas through the 'gene's eye view', in a non-technical piece,[4] and was ready to embrace the new more thoroughgoing analysis wholeheartedly.[5] The next adaptationist advance was evolutionarily stable strategies,[6] which were explained by a context-dependence in which the environment of a replicator is affected by the nature of other replicators in the vicinity. Another was Trivers' reciprocal altruism,[7] which would work only if a replicator could rely on the returned favour coming back precisely to itself, at least most of the time. The exposition of these and other ideas in *The Selfish Gene* is not only a tour de force of plain speaking but, by relating them all to the same central argument, itself the best representation of Darwinism available, it established a single conceptual framework within which old and new ideas in adaptationism could be understood.

This overarching coherent structure in *The Selfish Gene* provides the kind of logical foundation and conceptual unity across a broad spectrum of ideas that is usually associated with mathematics. The irony will become clear. First, I want to note that the successes of the book, popular and academic, stem not only from the clarity of exposition and beautiful use of language that are universally acknowledged in Dawkins' work, but also from the less appreciated, but in intellectual terms much more significant, fundamental contribution to science represented by this foundational structure.

It is time to turn to the question of why mathematics did not provide, and has still not provided, a unification of the adaptationist theories under a Darwinian umbrella. The basic answer is that the mathematical arguments are rather complicated, and need to link branches of mathematics in a way that no one at the time was doing. My own current research programme is exactly on these lines, but while there has been significant progress and the direction is clear, there is much still to do.[8]

Not only was a mathematical unifying theory not provided at the time, two factors made the reception of *The Selfish Gene* by mathematical biologists at best difficult. First, there was a clear sense that the 'only true church' was mathematical models of Mendelian genetics, and so anything in words must be suspect at least. Certainly, once the word 'selfish' was used, mathematicians would recoil from the use of (as they saw it) metaphor, and deny that purpose could be found in equations describing gene frequencies. I strongly suspect that this scepticism about verbal arguments applied also to Darwin's own arguments, and that few mathematical population geneticists in 1976 (or today?) would have admitted the validity of the arguments in the *Origin of Species* or the *Descent of Man*, or even R. A. Fisher's *Genetical Theory of Natural Selection*.[9]

The second and more particular factor requires a brief excursus into the history of evolutionary ideas. Darwin's verbal arguments convinced many biologists, but when Mendelian genetics was rediscovered in the first decade of the twentieth century, it was widely believed that Darwinism and Mendelism were in substantial conflict. It was one of the founding fathers of population genetics, R. A. Fisher, who made it his task to investigate whether and how Darwin's conclusions could be justified on the basis of Mendelian mechanisms of inheritance. In one of those papers any theoretician would give their eye-teeth to have written, Fisher[10] established both that there was no conflict, and that Mendelian genetics actually resolved the main outstanding difficulty at the time for Darwinism, which was how variation could be maintained in the face of the apparent blending of parental traits in the production of offspring. More directly relevant here is Fisher's 'fundamental theorem of natural selection', which he believed to have the same significance for biology as the second law of thermodynamics has for physics. This theorem encapsulates the link between changes in gene frequency and adaptation. It states that gene frequencies change in a way that increases the mean fitness of the individuals. I have stated this rather carefully and, noting that there are considerable sophistications here, I pass quickly on. What is germane here is that the theorem, first published in 1930,[11] seems to provide a unifying theory for Darwinism in terms of Mendelism, and so a natural starting point for a mathematical unification of the new theories of the adaptationist renaissance. The quantity to be maximized is an individual's number of offspring, and Fisher devotes a section of the 1958 edition of his classic work to arguing that the individual is the agent that will seem to maximize it.

The reaction of respecters of mathematical population genetics to *The Selfish Gene* has to be understood in terms of what had happened to the 'fundamental theorem' by 1970. The history is told by Edwards,[12] and I have contributed a few observations of my own.[13] Very briefly, the theorem was seriously misunderstood by everyone except Fisher. By 1970, population geneticists had comprehensively disproved an erroneous version of the theorem, and its reputation lay in tatters. The theorem was seen as wishful thinking towards general biological principles, exposed as fallacious by the rigours of modern mathematical methods. There was therefore a very inauspicious atmosphere for any general claims about adaptation, or fitness, or maximizing principles. 'Where are your equations?' was the not always implicit challenge, and those without equations altogether tended to be viewed as time-wasting hopeful simpletons. It is of interest that when Hamilton published his theory of the maximization of inclusive fitness in 1964,[14] he showed great sensitivity to this ambience, and tried not to draw attention to the close similarities between Fisher's result and his own. All the same, mathematical population geneticists were simply dismissive of Hamilton's claims.[15]

Before returning to *The Selfish Gene*, it is important to indicate that the misunderstanding of the fundamental theorem is now undone, and the theorem's truth (though not its significance) is today widely accepted.[16] My own work[17] has recently provided an account of Hamilton's theory that is consistent with the mathematical standards of population genetics. So the negative atmosphere was unnecessary and logically unjustified, but no less real at the time for that.

Imagine, then, reading *The Selfish Gene* as a biologist in 1976 and being persuaded by it. If you know little of population genetics, your response can be uncomplicated approval. If you do know something, or know someone who does, you will be aware that the basic known processes of natural selection consist of Mendelian genetics, and that gold-standard arguments about natural selection are mathematical models of genotype frequencies, and that the arguments in *The Selfish Gene* are therefore, well, what? Resolutions of this dilemma were various. One could embrace *The Selfish Gene*, and have faith that at some future time appropriate theoretical work would be done that reconciled it to existing Mendelian theory—in retrospect, this rose-tinted view seems the right one. Or one could hold on to the message by qualifying it in some vague way so that it needn't come into direct logical conflict with the mathematical models: for example, one could view it as an approximation, or an excellent explanation but at a superficial level. This qualified acceptance seems to have dominated the 'official' scientific response—it is noteworthy that Dawkins' election to the Royal Society was for his contribution to the public understanding of science, not for his contribution to science itself. Another response altogether was to reject the persuasive arguments, on the grounds that they conflicted with already known principles. This, as we've seen, turned out in retrospect to be unnecessary, but it was the fully logical response for a biologist who accepted what was then 'known' from mathematical population genetics.

In short, readers untrammelled by the authority of population genetics are likely to have accepted the arguments in *The Selfish Gene* at face value; those who accepted that authority fully would have rejected them; and a large number of biologists, caught in between, had to find a way to reconcile the conflict.

The apparent conflict with population genetics explains how the arguments in *The Selfish Gene* can have been right, but been rejected by theoreticians at the time. It explains why within biology the considerable scientific contributions it makes are seriously underestimated, and why it is viewed mainly as a work of exposition.

The fully logical arguments of *The Selfish Gene* give the book its enormous punch. A generation of biologists learnt about natural selection from *The Selfish Gene*. A conceptually unified view of natural selection is learnt with more intellectual satisfaction and more pleasure than a disparate one, which may well have contributed significantly to the huge expansion in the teaching of Darwinian biology in that era. The relevance of

Darwinian arguments to neighbouring disciplines becomes much clearer when a thoroughly thought through logical development is presented from first principles, and the extension of Darwinism into anthropology and evolutionary psychology, among other subjects, was greatly facilitated by the book. Within biology, too, the clarity of the vision offered by *The Selfish Gene* encouraged field biologists to apply the newfangled ideas they now understood, and theoreticians to take up and further develop the theories that now seemed so central to the concerns of biology.

I am convinced that *The Selfish Gene* brought about a silent and almost immediate revolution in biology. The explanations made so much sense, the fundamental arguments were so clearly stated and derived completely from first principles, that it was hard to see after reading the book how the world could ever have been any different. Indeed, for the many who learned about natural selection through reading *The Selfish Gene*, the world never had been any different. The very transparency of the exposition tended to make the book itself invisible within the newly created conceptual structure. A reason for another kind of academic invisibility is that, because of the authority of mathematical population genetics within biology, and the mistaken disapproval already alluded to, there was in many cases a sense that it wasn't respectable to cite *The Selfish Gene*.

The verbal and conceptual approach that we find equally in the *Origin of Species* and *The Selfish Gene* has its limits, but it is less often recognized that the mathematical approach also has limits. The limits to verbal argument come as ideas become more complex and more particular, when there are too many competing forces for the brain to hold them all in play at once; then formalism is required, and a sufficiently logical way of writing down intermediate results so that we know exactly what they mean and when they mean it. The limits to mathematics arise from the less formal, more substantive, question of the appropriateness of a mathematical framework for a biological problem. Perhaps the lesson here is simply not to trust a mathematician who dismisses a biological work, unless they can persuade you they've understood it first.

This brief sketch has omitted much for the sake of focusing on *The Selfish Gene* and its intellectual contribution to science. There were other verbal and conceptual thinkers, eminent among them G. C. Williams and R. L. Trivers (though Trivers could certainly do algebra when required),[18] but it was Dawkins who re-expressed the fundamental logic of Darwinism and brought all the existing theory into coherence with it. Dawkins went on to expand his Darwinian logic, with the replicator/vehicle distinction,[19] which has also gone on to be studied and developed by philosophers, and has contributed many further explanations of evolutionary topics. The link between the political objections to *The Selfish Gene* and the rejection of it by mathematical population geneticists is fascinating, but too large a topic to be included here. The mathematical arguments needed to provide formal structures to parallel Dawkins' conceptual structures are vital for establishing my claim that the rejection by mathematical population geneticists was erroneous, but cannot be given in this brief essay, and anyway are the subject of my continuing unfinished researches.[20]

What should not be omitted is a recommendation to reread *The Selfish Gene*. After thirty years, the logic is as impeccable, the explanations are as clear and as fresh, and the whole argument still draws the reader in inexorably. I have only mentioned the large-scale arguments so far in this article, the giants among the ideas, but it is also astonishing how lucid and absolutely right all the detailed arguments are, how the complexities are all recognized and deftly included. The message, then, of this essay, is that there is no need to feel reservations about the lack of mathematics, and that if there are sharply in-drawn breaths of imagined theoreticians at one's shoulder (as described in the book's original preface), they are unjustified. *The Selfish Gene* was a work of immense scientific creativity in 1976, providing the conceptual foundations and unifying framework of modern Darwinian biology, and remains unsurpassed, whether by word or by mathematics, to this day.

ENDNOTES

1 R. Dawkins, *The Selfish Gene* (Oxford: Oxford University Press, 1976).

2 G. C. Williams, *Adaptation and Natural Selection* (Princeton: Princeton University Press, 1966).

3 W. D. Hamilton, 'The genetical evolution of social behaviour', parts 1 and 2, *Journal of Theoretical Biology*, 7 (1964): 1–52.

4 W. D. Hamilton, 'The evolution of altruistic behavior', *American Naturalist*, 97 (1963): 354–356.

5 W. D. Hamilton, 'The Play by Nature (Review of *The Selfish Gene*),' *Science*, 196 (1977): 757–759.

6 J. Maynard Smith and G. R. Price, 'The logic of evolutionary conflict', *Nature*, 246 (1973): 15–18.

7 R. L. Trivers, 'The evolution of reciprocal altruism', *Quarterly Review of Biology*, 46 (1971): 35–57.

8 A. Grafen, 'A first formal link between the Price Equation and an optimization program', *Journal of Theoretical Biology*, 217 (2002): 75–91.

9 R. A. Fisher, *The Genetical Theory of Natural Selection* (Oxford: Oxford University Press, 1930; 2nd edn., Dover, 1958; 3rd variorum edn., Oxford University Press, 1999).

10 R. A. Fisher, 'The correlation between relatives on the supposition of Mendelian inheritance', *Transactions of the Royal Society of Edinburgh*, 52 (1918): 399–433.

11 Fisher, *The Genetical Theory of National Selection* (1930).

12 A. W. F. Edwards, 'The fundamental theorem of natural selection', *Biological Reviews*, 69 (1994): 443–474.

13 A. Grafen, 'Fisher the evolutionary biologist', *Journal of the Royal Statistical Society: Series D (The Statistican)*, 52 (2003): 319–329.

14 Hamilton, 'The genetical evolution of social behaviour' (1964).

15 A. Grafen, 'William Donald Hamilton', *Biographical Memoirs of Fellows of the Royal Society*, 50 (2004): 109–132.

16 See Edwards, 'The fundamental theorem of natural selection' (1994) and Grafen, 'Fisher the evolutionary biologist' (2003).

17 A. Grafen, 'The optimisation of inclusive fitness', *Journal of Theoretical Biology* (2005).

18 R. L. Trivers and H. Hare, 'Haplodiploidy and the evolution of the social insects', *Science*, 191 (1976): 249–263.

19 R. Dawkins, *The Extended Phenotype* (Oxford: W. H. Freeman, 1982).

20 See Grafen, 'A first formal link between the Price Equation and an optimization program' (2002) and 'The optimisation of inclusive fitness' (2005).

AN EYE ON THE CORE: DAWKINS AND SOCIOBIOLOGY

Ullica Segerstråle

'SOCIOBIOLOGY' is a term that triggers different reactions in different people. For some, sociobiology simply denotes a field devoted to research in animal social behavior, a new integrative discipline that emerged during the second part of the twentieth century. For others, because of its perceived implications for humans, this term is fraught with political connotations: genetic determinism, political conservatism, racism, and sexism. As a result, many researchers, especially those in would-be 'human sociobiology' are avoiding the term, calling themselves instead behavioral ecologists, evolutionary psychologists, Darwinian anthropologists, and what have you. For a third, more militant group, sociobiology is indeed a new discipline—a new scientific way of looking at behavior, maybe even a new world view, mobilizable against all kinds of obfuscation.

What lies behind this is the memory of the sociobiology debate, starting in 1975 around Harvard zoologist Edward O. Wilson's huge, popularly written tome, *Sociobiology: The New Synthesis*.[1] Wilson was being attacked by other scientists, including biologists from his own department, for misleading the general public and policy makers with bad and dangerous science. Over the years, the debate transformed itself into a trans-Atlantic affair and into a more scientific and less overtly political controversy (although the political undertones remained).[2] One scientist who was willy-nilly swept up in the debate at an early point was Richard Dawkins.

There are a lot of ironies involved in the story of Dawkins' involvement in the controversy—not least the fact that Dawkins did not regard himself as a sociobiologist (something that few people knew), and that the message of *The Selfish Gene*, published a year later, was largely different from Wilson's.[3] But in the prevailing climate dominated by the critics' political attacks on sociobiology, this did not matter. For the general public, at least in the United States, and for parts of the academic community as well, Wilson and Dawkins represented the same type of politically suspect 'bad' science. Sociobiology was nothing but old ideology masquerading as purportedly new science, stated the Sociobiology Study Group, the leading American group of critics associated with Science for the People. This group went as far as challenging the readers of *Science* to 'look for themselves' to find the obvious political messages of sociobiology.

In other words, whatever Wilson and Dawkins themselves as scientists had wanted to achieve with their books had been effectively subverted by the critics, who turned the issue into a moral and political one. The critics were able to make their interpretation of sociobiology prevail for a surprisingly long time. Under the mid-1970s' dominant 'environmentalist' or rather 'culturist' paradigm it was still too early to discuss any biological underpinnings of human behavior. That explanatory framework was not to lose its grip until near the end of the twentieth century.

What was it that Wilson and Dawkins actually wanted to convey with their popular books in the mid-1970s? Both wanted to present to a larger audience a synthesis of several new lines of theory and empirical studies of animal social behavior. Both emphasized the importance and fruitfulness of an evolutionary perspective for understanding behavior. Indeed, the whole idea of treating social behavior as something that was undergoing evolution—just in the same way as morphological traits—was a relative novelty in science at the time.

The cornerstone of the new line of reasoning had come in the early 1960s with Bill Hamilton's crucial insight into the evolution of altruism, that most selflessly social of behaviors, by which animals assist others even to the point of self-sacrifice. Darwin had regarded animal altruism as a major problem, 'one special difficulty, which at first appeared to me insuperable and actually fatal to my whole theory'. Why would an animal ever behave so as to reduce its biological fitness by putting itself in danger (say, letting out an alarm call to save other animals), or by foregoing reproduction (like the workers of many social insects)? This did not seem to give the animal any evolutionary advantage; quite the contrary. But Hamilton was able to show mathematically how this counterintuitive trait could actually spread. What is required is simply that the benefits of an altruistic act do not fall on random members of a population but on individuals that are genetically related to the donor.

In other words, Hamilton proposed to take a new look at fitness. It was not the fitness of the individual organism that counted, he reasoned, but the fitness of the whole group of relatives, because they also carry the individual's genes. The proportion of shared genes is relative to the closeness of relationship—for instance, on average one-half of the genes for full siblings are identical by common descent; the same goes for parent and child; the proportion is one-eighth for first cousins, and so on, making the respective 'coefficient of relationship' one-half, one-quarter, and so on). In this view, it does make evolutionary sense for an animal to risk its own life by saving a whole bunch of relatives— just how many and how closely related can be calculated by Hamilton's Rule. The rule says that for altruistic behavior to come about, the benefit (b) of an act has to outweigh its cost (c) times a number that is the reciprocal of the coefficient of relationship. This can be formulated as $b > 1/r \times c$ (this can be intuitively understood by referring to J. B. S. Haldane's London pub quip that he would willingly die for more than two brothers, four half-brothers, or eight cousins. It remained a quip, though, and Haldane did not develop the idea further into a universal rule).[4]

The term Hamilton invented for this new way of thinking was 'inclusive fitness'. John Maynard Smith's term 'kin selection' elegantly captured the basic idea of taking into account the whole cluster of gene-sharing relatives in fitness calculations. Nowadays 'kinship theory' is generally used to refer to Hamilton's insight.

Hamilton made his seminal contribution while he was still a graduate student in a long two-part paper ('The Genetical Evolution of Social Behavior', I and II) and a short paper summarizing his findings (1964 and 1963 respectively).[5] Many who read the long paper found especially the mathematics part hard going. This was a time when biologists were not routinely trained in mathematics or population genetics, despite the recent triumph of the Neo-Darwinian or Modern Synthesis, which reconceptualized the process of evolution mathematically as the change of gene frequencies in a population.[6] But others were able to read Hamilton's papers and early on realize their profound significance.

One of these was young Richard Dawkins. As a tutor in zoology at Oxford, he was well-trained in trying to convey difficult ideas to his students in easy-to-understand ways. (A student of the Nobel prizewinning ethologist Niko Tinbergen, he may have absorbed his mentor's credo that everything should be explained as clearly as possible.) Another was the Harvard entomologist Edward O. Wilson, who had taught himself population genetics, convinced that the Modern Synthesis had opened up promising new avenues for evolutionary biology. Both of these men felt the urge to share the new knowledge they had accumulated with a maximally wide public. The result was their popular works—Wilson's a large and heavy coffee-table book of over 500 pages, *Sociobiology: The New Synthesis* (published in the US by Harvard University Press) and Dawkins' more pocket-sized *The Selfish Gene*, a book of some 200 pages (published in the UK by Oxford University Press).

Many at the time assumed that Wilson's and Dawkins' books had the same general content. Some may even have believed that since *The Selfish Gene* came out after *Sociobiology*, it was influenced by the former. This is incorrect—these works were quite independent. This should have been obvious, too, to anyone who read both books. The question is, however, how many actually did—especially the huge and rather costly Wilson tome. It was easier to read *Sociobiology* 'through' the critics.

And the way the critics presented *Sociobiology*, it may have seemed to many that they had read a sufficient account of the book. Wilson was in fact described as putting in twenty-five chapters on animals as a sort of camouflage for his real message, supposedly contained in his first and last chapters on humans. The sociobiology controversy at least initially fixated itself on Wilson's final chapter, while the rest of the book was referred to as containing recent theories and empirical findings.

Now I come to my main concern. What were the 'socio-biological' syntheses of Wilson and Dawkins actually like? What were their core emphases? Which were the theories and recent empirical findings covered in *Sociobiology* and *The Selfish Gene*, respectively? What was each book's message? To anticipate, I shall argue that we are here dealing with two very different syntheses— indeed, two very different 'sociobiologies'—and that if one is looking for the actual core of new evolutionary ideas, the essence of the emerging new evolutionary paradigm, one had better con-sult *The Selfish Gene* rather than *Sociobiology*. However, if one desires a more comprehensive view, a wealth of contemporary theories and observations on animals in their environments—a naturalist's synthesis, as it were—then the obvious choice is *Sociobiology*. Let's look more closely at these alternative 'sociobiologies'.

Looking first at the aims of *Sociobiology* and *The Selfish Gene*, both aimed to introduce their readers to new discoveries in the field of evolutionary biology and a new way of thinking about social behavior. Wilson himself saw his book as an encyclo-pedia—and this it truly was, with its substantial extent and great number of references. But it was much more than an encyclopedia. It was partly Wilson's demonstration to his zoological colleagues at large that a common field of sociobiology actually existed across all specialist animal research, and that many common problems could now be analyzed in a new light. In other words, Wilson created, or codified, the field of 'sociobiology'—a feat gratefully recognized by the Animal Behavior Society, which in 1989 rated *Sociobiology* the most important book on animal behavior of all time.[7]

In synthesizing *Sociobiology*, Wilson very much remained in the naturalist tradition. In fact, many have observed that large parts of Wilson's big tome read like an ethology textbook. The material of the book is very rich on descriptive detail on animal behavior, and full with illustrations. Indeed, even in naming the field 'sociobiology', Wilson saw himself as following the tradition established by John Paul Scott, who coined the term to mean 'the interdisciplinary science which lies between the fields of biology (particularly ecology and physiology) and psychology and sociology'.[8] It later acquired a semi-official status within a subsection of what was to become the Animal Behavior Society.

But Wilson's own definition of 'sociobiology'—the systematic study of the biological basis of all social behavior—had implicitly a broader aim. Not only did Wilson want to show to his scattered fellow researchers that a common focus existed, he wanted to convey to a larger lay public as well a new integrative way of looking at behavior. His book was a plea for taking biology seriously in explaining humans, too—not exempting our species because of such things as language, culture, and learning. Accordingly, to demonstrate this point he provided bits and pieces of suggestive research in his last chapter. Wilson's larger concern was the future of humankind (as part of life on this planet): it was important to know human nature for adequate social planning. What he wanted was to start a discussion about human nature and the future of humankind.[9] What he got was the sociobiology debate.

This was a nasty debate. Accusations were flying, demonstrations were rampant, anti-sociobiology gatherings had the character of religious revival meetings. The high point of anti-sociobiological fervor was a pitcher of ice water being poured over Wilson's head as he prepared to speak at a sociobiology symposium at the American Association of the Advancement of Science.[10] The mid-1970s was not a good time for getting a hearing for biological factors in social behavior. (The United States had seen a similar attempt just a few years earlier with the Berkeley psychologist Arthur Jensen's suggestion that differences in average

IQ scores between black and white American school children might be due to genetic differences—Jensen among other things had got his car tires slashed.) Wilson may have thought he had successfully dealt with IQ and race in *Sociobiology*. What he didn't expect was that the very idea of evolutionary explanation of human behavior would stir up a storm.

The big political concern of the critics (largely 1960s academic radicals) was that any biological explanation would automatically legitimize the social status quo and discourage social reform. This belief went hand in hand with the assumption that bad social consequences inevitably come from bad science. Accordingly, to eliminate this danger, the critics appointed themselves the task of weeding out anything that seemed bad. Up by the roots came sociobiology, IQ research, and behavioral genetics—anything involving modeling and indirect methods (not surprisingly, many leading critics came from experimental fields in science). A critical industry was born, and with it a flurry of articles revealing the purported hidden ideology, racism, sexism, and so forth in 'bad science'.[11]

To return to Wilson and his own intent with his book, we get one more clue on the very first page, where he presents altruism as the central problem of sociobiology. Aha! This looks like something that might lead to Hamilton. But Wilson uses this opening about altruism to make a philosophical point: to explain why human suicide—contra Camus—is in fact biologically informed rather than a product of free will.[12] In Chapter 1 we learn of yet another aim of his book: a bid for an even larger synthesis, the unification of the social and natural sciences (a goal later pursued in *Consilience*, 1998).[13] Wilson hopes that the scientific explanations of sociobiology will soon be able to replace the unscientific ones of the social sciences and ethics. But Wilson also has a more specific formulation of what he considers the goal of sociobiological explanation:

The ultimate goal is a stoichiometry of social evolution. When perfected, the stoichiometry will consist of an interlocking set of models

that permit the quantitative prediction of the qualities of social organization—group size, age composition, and mode of organization,
including communication, division of labor, and time budgets—from a
knowledge of the prime movers of social evolution.[14]

(The 'prime movers' here are phylogenetic inertia (resistance to
change) and ecological pressure.) In this version Wilson's sociobiology is an attempt to make evolutionary biology a total quantitative and predictive science.

Wilson's sociobiology, then, is at the same time an umbrella
discipline encompassing a wide range of studies of social behavior;
a more specific scientific program, employing particular explanatory tools; and a philosophical program for integrating the social
and natural sciences and ethics. Clearly in outlining his New
Synthesis, Wilson attempted nothing less than a revolution.

Looking back at *Sociobiology*, there is no doubt that it had
great impact—above all as a comprehensive, comparative view
of the genetic foundations of social behavior, and as a breaker
of the prevailing taboo on biological explanations of humans. But
to what extent did Wilson's *Sociobiology* convey the core ideas of
the paradigm shift in evolutionary biology that in reality
occurred?

Let us examine some of the core ingredients in the mid-1970s
making up the emerging evolutionary paradigm. Certain new
ideas and a new way of thinking appeared and excited scientists,
even to the extent that some experienced having undergone a
personal paradigm shift. The central idea would have to be
Hamilton's Rule and his concept of inclusive fitness, or alter-
natively, kin selection (for many, the personal paradigm shift
happened exactly through abandoning group selection for kin
selection explanation).[15] Then there are Robert Trivers' crucial
papers on reciprocal altruism, parental investment, and parent–
offspring conflict, and John Maynard Smith's Evolutionarily
Stable Strategy.[16] Beyond these we have George Williams' famous
dictum (or, as he himself called it, 'doctrine'), effectively ruling
out group level explanations in evolutionary biology:

The ground rule—or perhaps doctrine would be a better term—is that adaptation is a special and onerous concept that should be used only when it is really necessary. When it must be recognized, it should be attributed to no higher level of organization than is demanded by the evidence. In explaining adaptation, one should assume the adequacy of the simplest form of natural selection, that of alternative alleles in Mendelian populations, unless the evidence clearly shows that this theory does not suffice.[17]

This type of population genetic consideration of gene (allele) frequencies within populations drew attention away from the group and the individual to a new type of gene-selectionist thinking. What emerged was a fundamentally game-theoretical way of approaching behavior where individuals (affected by their individual genes) appeared as strategists, calculating as it were the best way to behave in order to further their own inclusive fitness. Gone were especially the vague post-World War II explanations that individuals sacrifice themselves 'for the good of the species'; in were clear-cut testable mathematical models of what an animal in a particular situation would do, depending on the actions of other animals present. With this core in place, the gates were opened for a whole new scientific industry, turning out to be tremendously successful.[18]

Looking now at *Sociobiology* as a source of information about the emerging paradigm, it is clear that this book, despite its other merits, was *not* an attempt to convey the gist of the new way of thinking outlined above. It represented a broader, more comprehensive synthesis, with its own, different goal. Although many of the crucial contributors are mentioned in *Sociobiology*, they are not presented as being somehow connected to one another or as representing a common perspective—in fact, they and their insights become easily lost in the mass of other names and the overall richness of information.

Also, when we look at how Wilson treats some of the key theorists, he sounds surprisingly unenthusiastic. Take Bill Hamilton. Wilson doesn't mention Hamilton and his crucial contribution until Chapter 5 in *Sociobiology*. And there the theory of kin selection is introduced as only one of many group selectionist theories explaining altruism, not as any revolutionary new discovery. In other words, in his book Wilson is *not* promoting a move away from 1960s 'good for the species' group selection in favor of kin selectionist thinking: he is, if anything, doing the reverse! Although Wilson does present Hamilton's ideas in a big diagram in this chapter, he appears rather ambivalent about them. For Wilson's taste, Hamilton's approach is too limited.[19]

What about George Williams, whose 1966 book *Adaptation and Natural Selection* many saw as so central? Williams is not a guru for Wilson. Wilson waves away Williams' influential book already on the second page of his book. What is wrong with Williams? Williams is not group selectionist enough![20]

How does Wilson treat Robert Trivers? Trivers' ideas get fair coverage: in their own right, however, rather than as harbingers of a new way of thinking. Trivers' important new concept of reciprocal altruism appears in Wilson's Chapter 5. But Trivers' other core ideas of parental investment, and parent–offspring conflict appear well after page 300 in *Sociobiology*.

Finally, let's check John Maynard Smith. His important concept of Evolutionarily Stable Strategy (ESS) is not mentioned in *Sociobiology* at all.[21] Game theory and strategies are not organizing concepts in Wilson's *Sociobiology*.

If we go to Dawkins' *The Selfish Gene*, on the other hand, all of the crucial components and all the crucial theorists are to be found there, and enthusiastically presented. *The Selfish Gene* is an attempt to convey the underlying logic of a particular type of reasoning, rather than representing a broad overview. Examples are brought in as illustrations, some of these quite hypothetical. Dawkins wants to present the unifying core of the new theoretical contributions. What do the different new approaches have in common, how are they connected? The common umbrella is clearly game theory and gene selectionist thinking. But Dawkins goes a step further. Since it is the transmission of genes that matter in evolution, why not follow the genes around and see what kinds of strategies they seemingly use? We know the interest of the genes: to get themselves or copies of themselves into the next generation. Developing the gene's eye view as a pedagogical tool for understanding such concepts as inclusive fitness and parental investment, Dawkins provides a window also on population genetics. In the process Dawkins' heuristic device becomes a conceptual glue for keeping the different core ideas of sociobiology together.

I have here presented Wilson and Dawkins as very different. And that is also how they themselves experienced the situation. In fact, in an interview with me in the early 1980s Wilson considered Dawkins as having made 'a mistake' with his notion of the selfish gene. Here Wilson joined his older Harvard colleague Ernst Mayr who considered it completely incorrect to speak of genes—one should only speak about genotypes. (Surprisingly for some, perhaps, who followed the conflict between Wilson, Gould, and Lewontin, these three together with Mayr can be classified as representing the would-be Harvard school in evolutionary biology—a school that did not accept genes, only genotypes, and that collectively regarded the Dawkinsian type of sociobiology as anathema.)

For many reasons, then, it is hard to regard Wilson's sociobiology as being similar to Dawkins'. However, from a political analyst's point of view, the mere fact that both scientists were using genetic explanations for social behavior was enough. It did not matter to the critics either that, unlike Wilson, Dawkins did not try to include humans—humans appear in a separate last chapter, where memes (units of culture), not genes, are king. The political critics, set on finding fault with the book, blatantly ignored Dawkins' disclaimer that he with his title was describing a new way of looking at evolution, not exhorting humans to be selfish. In relation to both *Sociobiology* and *The Selfish Gene* the critics employed a particular reading strategy which always yielded results (I have called this 'moral reading').[22] The aim was to imagine the worst possible social or moral implications of selected sentences in the book. These then justified their condemnation.

While the American Science for the People were quite good at this (with Wilson's colleagues Gould and Lewontin among the initiators of the fault-finding effort), Dawkins' particular nemesis was Britain's Steven Rose. Rose enjoyed taking abstract statements about strategies from Dawkins, applying them to humans and then condemning the author for justifying socially unacceptable behaviors (Dawkins once characterized Rose's misreading of him as 'a wanton will to misunderstand'). For his American critics with their peculiar text reading habits, Dawkins' vivid language and colorful illustrations of genetic strategies were also a gold mine of shocking statements when his examples were taken literally.[23]

I have tried to show that whatever *Sociobiology: The New Synthesis* was a synthesis of, it was not a synthesis of the core new ideas of inclusive fitness, Evolutionarily Stable Strategy, parental investment, and so on. Wilson did not himself accept a gene-selectionist approach or work within a game-theoretical explanatory framework. It is perhaps strange, then, that many take for granted that it was Wilson's 1975 book that conveyed the basic ideas of sociobiology, as these were later commonly understood— an aggregate of a certain set of ideas united by a gene-selectionist and game-theoretical framework.

One reason was clearly the critics' tendency to lump together Wilson's and Dawkins' sociobiology under a common political umbrella. It is also possible that the critics, who read mostly the last chapter of *Sociobiology*, just believed that the book presented the same paradigm as Dawkins'. Moreover, there was a type of broader philosophical criticism that tended to lump the different types of sociobiology together. For instance, from the very beginning the critics accused both Wilson and Dawkins of various types of 'error', especially for being 'reductionists'.[24]

Another widespread impression is that it was Wilson's *Sociobiology* that finally gave Hamilton's important ideas their deserved publicity. (Hamilton's revolutionary 1964 paper was not widely recognized or cited in the literature until the mid-1970s.[25]) The role of *Sociobiology* in promoting Hamilton may well be something of a charming myth, considering this book's actual presentation of his contribution.[26] In order to pay special attention to Hamilton among the mass of information in *Sociobiology*, a reader would probably have to know about him already from somewhere else.

Dawkins, on the other hand, has no hesitation about Hamilton being the core of sociobiology. For him sociobiology is in fact 'the branch of ethology inspired by Bill Hamilton'.[27] Accordingly, *The Selfish Gene* expounds on such things as The Prisoner's Dilemma as the prototype model for game theoretical reasoning. It teaches the reader to start thinking in terms of strategies, and it provides a common conceptual framework for the core theorists Hamilton, Williams, John Maynard Smith, and Trivers. The book is full of imaginative examples, some of them involving genetic actors as vivid as if they were humans, all in the service of explaining the logic or mechanism of evolution, and all from a gene's eye perspective.

It fell on Dawkins to be the one to clear up misunderstandings about sociobiology. Wilson had rather soon after *Sociobiology* moved on to gene-culture coevolutionary models and later on to preserving biodiversity. After the early 1980s, there was little visible interchange between Wilson and his original critics. It was now Gould and Dawkins, rather than Wilson and Lewontin, who slowly emerged as the new 'public' scientists in the sociobiology controversy (Gould supported in the background by Lewontin, and Dawkins defended particularly by John Maynard Smith). At the same time, the controversy began increasingly to address important issues in evolutionary theory (without losing its political underpinnings).

The Extended Phenotype, Dawkins' second book,[28] was one huge attempt to catch and respond to general accusations and misperceptions when it came to such things as adaptation, unit of selection, the relationship between macro- and micro-evolution, and others. He tirelessly pointed out that using a gene selectionist framework was not tantamount to genetic determinism (myopia can be overcome with the help of glasses) and that using an adaptationist framework did not necessarily mean a commitment to a view of the best of all possible worlds (the accusation of Gould and Lewontin against 'adaptationists'). Elsewhere he cleared up twelve misunderstandings of kin selection.[29] He explained to the philosopher Mary Midgley that he used Chicago gangsters in order to illustrate a certain gene strategy, not to legitimize gangster behavior.[30] He showed in various ways how adaptation was actually able to produce complex traits. A later book, *The Blind Watchmaker*, was largely a response to Gould's challenges to adaptationism at the time: punctuated equilibria and developmental constraints.[31] And so Dawkins toiled on against the critics, in a stream of books and articles over the years.

In 1982 Dawkins had tried to clarify matters with the help of logic. Here is a typical patient response. The critics have got it wrong, Dawkins argued. The issue did not concern genetic determinism but genetic selectionism:

Gene selectionism, which is a way of talking about evolution, is mistaken for genetic determinism, which is a point of view about development. People like me are constantly postulating genes 'for' this and genes for that. We give the impression of being obsessed with genes and with 'genetically programmed' behaviour. . . .

Why, then, do functional ethologists talk about genes so much? Because we are interested in natural selection, and natural selection is differential survival of genes. If we are to so much as discuss the possibility of a behaviour pattern's evolving by natural selection, we have to postulate genetic variation with respect to the tendency or capacity to perform that behaviour pattern.[32]

But three years later Dawkins had had enough. The book that transformed Dawkins into the leading defender of the core of sociobiology was *Not In Our Genes* by Kamin, Rose, and Lewontin.[33] In his review of this book he stated that 'much as I have always disliked the name [sociobiology], this book finally provokes me to stand up and be counted'.[34] With this, Dawkins created for himself an opportunity to confront directly one of the critics' persistent claims about sociobiology:

Sociobiology is a reductionist, biological determinist explanation of human existence. Its adherents claim, first, that the details of present and past social arrangements are the inevitable manifestations of the specific actions of genes.[35]

At this point Dawkins exploded:

Rose et al. cannot substantiate their allegation about sociobiologists believing in inevitable genetic determination, because the allegation is a simple lie. The myth of the 'inevitability' of genetic effects has nothing whatever to do with sociobiology, and has everything to do with Rose et al.'s paranoiac and demonological theology of science.[36]

The authors of *Not In Our Genes*, however, did not let this pass as an understandable protest by a sociobiologist finally speaking out. Incredibly, Rose threatened to sue Dawkins for libel! Nothing came of the suit, so all calmed down.[37]

One effect of the constant criticism of sociobiology was that Dawkins felt forced to explain himself even better. At an early point he introduced the idea of the difference between replicators (genes) and vehicles (temporary bodies that carry the genes). Later on he clarified how one can have a gene's eye point of view and still accept the reality of different levels of selection. In many ways he succeeded in undermining the criticism launched against sociobiology, making it hard for his critics. (Politically too, as for instance when he mentioned in the second edition of *The Selfish Gene* that he voted Labour!)

But this did not seem to satisfy Dawkins' main critic, Stephen Jay Gould. The individual fight betweeen Gould and Dawkins continued. In 1995 Gould presented Dawkins as a 'strict Darwinian zealot',

who's convinced that everything out there is adaptive and a function of genes struggling. That's just plain wrong, for a whole variety of complex reasons. There's gene-level selection, but there's also organism-level and species-level . . .[38]

To this Dawkins responded:

The 'pluralist' view of evolution is a misunderstanding of the distinction I make between replicators and vehicles . . . There's a hierarchy in levels of selection as long as you are talking about vehicles. But if you are talking about replicators, there isn't. There's only one replicator we know of, unless you count memes.

Steve doesn't understand this. He keeps going on about hierarchies as though the gene is at the bottom level in the hierarchy. The gene has nothing to do with the bottom level in the hierarchy. It's out to one side.[39]

Later there emerged a more inclusive opposition between two larger camps defined by Gould and Eldredge. On one side were the 'pluralists' or 'naturalists', on the other the terrible 'ultra-Darwinians', which by now included both sociobiologists and evolutionary psychologists. Gould and Eldredge (pluralists and naturalists) were concerned with a comprehensive—and therefore 'correct'—explanation of evolution that would take into account all the forces and all the levels involved.[40] 'Theirs is an incomplete description of biotic nature, rendering their theory simplistic and incomplete', said Eldredge about the adaptationist ultra-Darwinians.[41] For him, what was required was the total truth about nature.

We have, therefore, a more general opposition between what we might more appropriately call ontological truth-seekers and those who are concerned with the mechanism of natural selection, mechanism-oriented truth-seekers, or 'logicians'. It is the difference between those who believe that evolutionary biology should answer 'Why?' questions and those who think that this science, just like the rest of science, should limit itself to 'How?' questions. This opposition represents a tension between an earlier and later tradition in evolutionary biology, but is also a matter of background training and personal taste. At the same time, this mirrors an earlier conflict among the architects of the Modern Synthesis.[42] The political aspects of the sociobiology debate have obscured the existence of this continuing tension between the 'Why?' and the 'How?' which also underlies the different conceptions of sociobiology of Wilson and Dawkins.

The central theme of this essay has been the profound difference between the Wilsonian and the Dawkinsian approach. Wilson's sociobiology was of a systemic, integrative kind. He was interested in bringing all social behavior under one big umbrella, making sociobiology a comprehensive quantitative and predictive science. Dawkins, in contrast, wanted to explain to his readers a new way of thinking about evolution. Dawkins' aim was to identify and convey the logic underlying a certain cluster of new evolutionary theories. The core of sociobiology (as we now know it) was articulated in *The Selfish Gene* rather than in Wilson's *Sociobiology*. Wilson's book was called 'the new synthesis', but for practicing sociobiologists, the ideas presented in Dawkins' book became the synthesis-in-use. And the concept that helped delineate and solidify the new sociobiological paradigm was the gene's eye view.

ENDNOTES

1 Edward O. Wilson, *Sociobiology: The New Synthesis* (Harvard, MA: Harvard University Press, 1975).
2 The sociobiology controversy and its broader scientific and political context are described in Ullica Segerstråle, *Defenders of the Truth* (Oxford and New York: Oxford University Press, 2000). This book, based largely on interviews with the main actors and their colleagues, follows the debate from 1975 to 2000.
3 Richard Dawkins, *The Selfish Gene* (Oxford and New York: Oxford University Press, 1976).
4 Sacrificing one's life is not the only way to altruism. Another is foregoing reproduction, as is the case with the worker caste in many social insects. One of Hamilton's prime examples of the workings of his rule was in fact the explanation why it makes sense for the workers of Hymenoptera to help raise the queen's offspring instead of their own. The reason is the unusual coefficient of relationship (3/4) that arises in certain Hymenopteran species between the queen and workers, because they are sisters under haplodiploidy. In such a species the workers are more closely related to the queen (3/4) than to their own daughters (1/2). (Because of the vividness of this example, many initially came to believe that the idea of kin selection applied especially, or exclusively, to Hymenoptera.) As for Haldane, according to John

Maynard Smith, 'there is no reason to think that Haldane thought the idea more than entertaining'. He goes on to say that Haldane did publish a version of this kin aphorism 'as a throw-away paragraph in a popular journal' (*New Biology*, 18 (1955): 34), but never mentioned it in a lecture, and never followed it up (Maynard Smith, letter to Bill Hamilton, 14 November 1980).

5 William D. Hamilton, 'The genetical theory of social behavior', I and II, *Journal of Theoretical Biology*, 7 (1964): 1–16; 17–32, and 'The evolution of altruistic behavior', *The American Naturalist*, 97 (1963): 354–356.

6 For a discussion of Neo-Darwinism and the Neo-Darwinian or Modern Synthesis, see e.g. Ullica Segerstråle, 'Neo-Darwinism', in Mark Pagel (ed.), *Encyclopedia of Evolution* (Oxford: Oxford University Press, 2002), 107–110. The synthesis took place in two steps. First there was the unification of Darwinism and Mendelianism in the 1920s and 1930s, under R. A. Fisher, J. B. S. Haldane, and Sewall Wright (whereby evolutionary principles were rewritten in the language of population genetics). Later the synthesis was expanded to other fields of biology, such as systematics, paleontology, and even botany. Some of the chief architects of this second stage were Ernst Mayr, Theodosius Dobzhansky, and Gaylord Simpson. The synthesis was largely finished in 1950, but not everybody was happy with the results (see e.g. Niles Eldredge, *The Unfinished Synthesis* (New York: Oxford University Press, 1985).

7 Edward O. Wilson, *Naturalist* (Washington, DC: Island Press, 1994), 330.

8 Edward O. Wilson, 'A consideration of the genetic foundation of human behavior', in G. W. Barlow and J. Silverberg (eds.), *Sociobiology: Beyond Nature/Nurture?* AAAS Selected Symposium 35 (Boulder, CO: Westview Press, 1980), 295–306.

9 See Segerstråle, *Defenders of the Truth* (2000), 365 ff.

10 See Segerstråle, *Defenders of the Truth* (2000), 23–24.

11 See Segerstråle, Defenders of the Truth (2000), Chapters 10 and 11.

12 Wilson starts his book *Sociobiology* as follows:

Camus said that the only serious philosophical question is suicide. That is wrong even in the strict sense intended . . . Self-existence, or the suicide that terminates it, is not the central question of philosophy. The hypothalamic-limbic complex automatically denies such logical reduction by countering it with feelings of guilt and altruism. In this one way the philosopher's own emotional control centers are wiser than his solipsist consciousness, 'knowing' that in evolutionary time the individual organism counts for almost nothing.

13 Edward O. Wilson, *Consilience: The Unity of Knowledge* (New York: Alfred Knopf, 1998).

14 Wilson, *Sociobiology: The New Synthesis* (1975), 63.

15 Hamilton's Rule is to be found in Hamilton 'The evolution of altruistic behavior' (1963). Kin selection was seen as the answer to the problem of altruism, which previous explanations in terms of group selection had not been able to resolve. In Maynard Smith's authoritative presentation, kin selectionist explanation became the alternative to the then prevailing group selectionist paradigm. Maynard Smith regarded group selection (selection between groups of altruistic individuals, individuals in each group acting 'for the good of the species') as a possible but unlikely phenomenon: altruists within a group would not have the chance to band together before getting outcompeted by selfish group members. Dawkins' *The Selfish Gene* also reflects this view of kin selection replacing group selection. This view was taken for granted in the new gene-selectionist paradigm emerging triumphant in the wake of post-war 'good for the species' talk. (Ironically, and seldom recognized, is that in the 1970s Hamilton himself came to regard kin and group selection not as antithetical, but rather as part of a continuum; see especially William D. Hamilton, 'Innate social aptitudes of man: An approach from evolutionary genetics', in R. Fox (ed.), *Biosocial Anthropology* (New York: John Wiley & Sons, 1975), 133–157. For later attempts to resurrect group selectionist thinking, see e.g. Elliot Sober and David Sloan Wilson, *Unto Others*. (Cambridge, MA: Harvard University Press, 1998).

16 Robert L. Trivers, 'The evolution of reciprocal altruism', *Quarterly Review of Biology*, 46 (1971): 35–57; Robert L. Trivers, 'Parental investment and sexual selection', in B. Campbell (ed.), *Sexual Selection and the Descent of Man* (Hawthorne, NY: Aldine, 1972); Robert L. Trivers, 'Parent–offspring conflict', *American Zoologist*, 14 (1974): 249–264; John Maynard Smith and George Price, 'The logic of animal conflict', *Nature*, 246 (1973): 15–18.

17 George C. Williams, *Adaptation and Natural Selection* (Princeton, NJ: Princeton University Press, 1966).

18 This is a good example of how a new way of thinking can open up a whole new line of research, something that has been recognized even by contemporary critics of the gene-selectionist paradigm. For instance, according to Niles Eldredge in 1995, Dawkins had been

'enormously good for the profession' and gene-selectionism started a whole 'kitchen industry' which 'gave lots of people lots of work' (Niles Eldredge, 'A battle of words', in J. Brockman (ed.), *The Third Culture* (New York: Simon & Schuster, 1995), 119–125.

19 For a discussion of group selection versus kin selection, see note 15.

Wilson says the following about Hamilton's idea of kin selection (or inclusive fitness):

The Hamilton models are beguiling in part because of their transparency and heuristic value. The coefficient of relationship, r, translates easily into 'blood', and the human mind, already sophisticated in the intuitive calculus of blood ties and proportionate altruism races to apply the concept of inclusive fitness to a re-evaluation of its own social impulses. But the Hamilton viewpoint is also unstructured. The conventional parameters of population genetics, allele frequencies, mutation rates, epistasis, migration, group size, and so forth, are mostly omitted from the equations. As a result, Hamilton's mode of reasoning can be only loosely coupled with the remainder of genetic theory, and the number of predictions it can make is unnecessarily limited.

From Wilson, *Sociobiology: The New Synthesis* (1975), 119–120.

20 According to Wilson:

Williams' Canon was a healthy reaction to the excesses of explanation invoking group selection and higher social structure in populations . . . Nevertheless, Williams' distaste for group-selection hypotheses wrongly led him to urge the loading of the dice in favor of individual selection. As we shall see in chapter 5, group selection and higher levels of organization, however intuitively improbable they may seem, are at least theoretically possible under a wide range of conditions. The goal of investigation should not be to advocate the simplest explanation, but rather to enumerate all of the possible explanations, improbable as well as likely, and then to devise tests to eliminate some of them.

For Wilson, the goal in sociobiology was to explain the complex mechanisms of behavior as *correctly*, not as simply, as possible. According to him, Williams' recommendation for constructing evolutionary hypotheses was just 'a more sophisticated variant' of what Wilson had called 'The Fallacy of Simplifying the Cause', a problem in sociobiological reasoning (Wilson, *Sociobiology*, 30).

21 An ESS is a 'strategy' (that is, a pattern of behavior) that is evolutionarily stable, which means that this pattern will prevail against any

alternative pattern when it is the dominant one in a population. Natural selection tends to produce populations that are evolutionarily stable. In practice, this often implies a particular balance of gene frequencies.

22 Segerstråle, *Defenders of the Truth* (2000), Chapter 10.

23 See e.g. Richard Dawkins, *The Extended Phenotype* (Oxford and San Francisco: W. H. Freeman, 1982), 10.

24 See e.g. Ullica Segerstråle, 'Reductionism, "Bad Science" and Politics: A Critique of Anti-Reductionist Reasoning', *Politics and the Life Sciences*, 11/2 (1992): 199–214.

25 The impression that it was *Sociobiology* that launched Hamilton was seemingly supported by a study by Paul Harvey and Jon Seger in 1980 (Jon Seger and Paul Harvey, 'The evolution of the genetical theory of social behaviour', *New Scientist* 87/1208 (1980): 50–51. Looking in the Science Citation Index, they discovered a 'mutant' in the citations to Hamilton's paper, appearing at a particular point. Many citers had used the formulation 'The Genetical Theory of Social Behavior' rather than the correct 'The Genetical Evolution of Social Behavior' when referring to Hamilton. Harvey and Seger traced the mutant reference to *Sociobiology*, where it indeed appeared in Wilson's bibliography. However, an examination of *The Selfish Gene* shows that this book, too, had the same incorrect mutant reference!

Moreover, it can be shown that a visible upswing actually started already in the early 1970s before the advent of either book, suggesting that someone else, perhaps Trivers—or perhaps Hamilton himself— may have been responsible for this (Hamilton did produce a number of short, visible papers in *Science* and *Nature* in the late 1960s and early 1970s). See more discussion in Segerstråle, *Defenders of the Truth* (2000), Chapter 5 and, by Dawkins, in *The Selfish Gene* (Oxford: Oxford University Press, 2nd edn., 1989).

But if we concentrate on the period 1975–1976, a good explanation may still be that, whatever their original source of Hamiltonian inspiration, when looking up the exact reference people simply chose to consult the biggest book around—Wilson's *Sociobiology*—or the handiest one—Dawkins' *The Selfish Gene*—instead of going to the library! Clearly, then, both Wilson and Dawkins can be suspected for having substantially contributed to the upswing in Hamilton's citation pattern. However, as I have indicated, looking at the text of the two books, it appears that it was Dawkins rather than Wilson who explicitly in writing boosted Hamilton in the mid-1970s.

26 As we saw, Hamilton's theory was just one among the many other theories explaining altruistic behavior in Wilson's Chapter 5 and it is not even clear that Wilson in *Sociobiology* is very enthusiastic about Hamilton, compared with other group-selectionist theories of altruism; see note 19.

27 Dawkins saw both of them as 'functional ethologists'. According to Tinbergen, ethologists are in the business of answering four equally important, different questions about behavior: its (proximate) causation, its development, its evolutionary history, and its (adaptive) function. Dawkins, just like Hamilton and Wilson, decided to concentrate on the last of Tinbergen's famous 'four questions'.

28 Richard Dawkins, *The Extended Phenotype: The Gene as Unit of Selection* (Oxford and San Francisco: W. H. Freeman, 1982).

29 Richard Dawkins, 'Twelve misunderstandings of kin selection', *Zeitschrift fur Tierpsychologie*, 51 (1979b): 184–200.

30 Richard Dawkins, 'In defence of selfish genes', *Philosophy*, October (1981a): 562–579.

31 Richard Dawkins, *The Blind Watchmaker* (New York: W. W. Norton, 1987).

32 Dawkins, *The Extended Phenotype* (1982), 19.

33 Richard C. Lewontin, Steven Rose, and Leon Kamin, *Not in Our Genes* (New York: Pantheon Books, 1984).

34 Richard Dawkins, 'Sociobiology: The debate continues', Book review: *Not In Our Genes*, *New Scientist* (24 January 1985): 59–60.

35 Lewontin, Rose, and Kamin, *Not in Our Genes* (1984), 236.

36 Dawkins, 'Sociobiology: The debate continues' (1985), note 34, page 59.

37 This I heard from Patrick Bateson and Richard Dawkins at the time. The threat of a libel suit caused a flurry of activity. I got almost implicated myself—Wilson told me that Bill Hamilton had contacted him about a copy of my dissertation, which, it was hoped, might contain some material that could be used in a possible legal defense. (Ullica Segerstråle, *Whose Truth Shall Prevail? Moral and Scientific Concerns in the Sociobiology Debate*, Harvard University: Department of Sociology, 1983). And Dawkins, one of my interviewees, wrote me a letter in 1986 asking (perhaps rhetorically) whether I could find a single political message in the writings of sociobiologists. This was no problem at all, since I was used to putting on the critics' hat and following their 'moral reading' of texts. I sent Dawkins a whole list of passages in Wilson's *On Human Nature* (Cambridge,

MA: Harvard University Press, 1978), which I knew either had been or easily could be interpreted as political statements by the critics.

38 S. J. Gould, 'The pattern of life's history', in J. Brockman (ed.), *The Third Culture* (New York: Simon & Schuster, 1995), 51–64.

39 Richard Dawkins, 'A survival machine', in J. Brockman (ed.), *The Third Culture* (New York: Simon & Schuster, 1995), 84.

40 This reminds us of Wilson's warning in *Sociobiology* to avoid 'The Fallacy of Simplifying the Cause', see note 20.

41 Eldredge, 'A battle of words' (1995), 122.

42 See Segerstråle, *Defenders of the Truth* (2000), 325 ff.

LOGIC

THE SELFISH GENE AS A PHILOSOPHICAL ESSAY

Daniel C. Dennett

One critic complained that my argument was 'philosophical', as though that was sufficient condemnation. Philosophical or not, the fact is that neither he nor anybody else has found any flaw in what I said. And 'in principle' arguments such as mine, far from being irrelevant to the real world, can be *more* powerful than arguments based on particular factual research. My reasoning, if it is correct, tells us something important about life everywhere in the universe. Laboratory and field research can tell us only about life as we have sampled it here[1].

PROBABLY most scientists would shudder at the prospect of having a work of theirs described as a philosophical treatise. 'You really know how to hurt a guy! Why don't you just say you disagree with my theory instead of insulting me?' But Richard Dawkins knows better. He is just as leery of idle armchair speculation and hypersnickety logic-chopping as any hardbitten chemist or microbiologist, but he also appreciates, as the passage above makes admirably clear, that the conceptual resources of science need to be rigorously examined and vividly articulated before genuine understanding, sharable by scientists and laypeople alike, can be achieved. Dawkins' contribution on this conceptual front is philosophy at its best, informed by a wealth of empirical work and alert to the way subtle differences in expression can either trap a thinker in an artifactual cul-de-sac or open up new vistas of implications heretofore only dimly imagined. My high opinion of his philosophical method is hard for me to separate, of course, from my deep agreement with the conclusions and proposals he arrives at. But this is not inevitable; there are few experiences more unsettling to a philosopher than watching a non-philosopher stumble into agreement with one's most carefully executed conclusions by a sort of lucky drunkard's walk. Dawkins, in contrast, is impressively surefooted.

I didn't read *The Selfish Gene* when it came out in 1976 because of some negative comment I ran into—I can't recall from whom—to the effect that the book was too clever by half, a bit of popularizing that could well be ignored. So I am deeply grateful to Douglas Hofstadter for undoing the damage of that bum steer fairly soon, in 1980, when he and I were working on our anthology, *The Mind's I*,[2] in which we included two excerpts, under the title 'Selfish Genes and Selfish Memes'. (Several times in my life I've taken the word of somebody I regarded well and moved a new book onto my 'don't bother' list only to discover later that this was a book that properly belonged on my 'read immediately' pile. We are all overwhelmed with competitors for our limited attention, so we really have no choice but to trust some filters and hope for the best, but it is distressing to find in retrospect that we have almost missed a close encounter of the finest kind. Ever since then I have tried—with marginal results, I'm sure—not to be dismissive of books unless I am really sure that they are time-wasters.)

I was a committed Darwinian before I got around to reading *The Selfish Gene*. My 1969 book *Content and Consciousness*,[3] and my essays, 'Intentional Systems',[4] and 'Why the Law of Effect Will not Go Away',[5] have Darwinian moves at their heart, for instance. But I actually knew very little about the fine points of the theory, and some of what I thought I knew was just wrong. *The Selfish Gene* delighted me from beginning to end, instructing and correcting me on dozens or hundreds of important points and confirming my inchoate sense that evolution by natural selection was the key to solving most of the philosophical problems I was interested in. This was mind candy of the highest quality. I have never looked back, as one says, so when I was invited to write an essay for this volume, it struck me that looking back would indeed be a good idea. Although I had often assigned large parts of the book in my classes, I hadn't reread it in one go, and now that I had spent a solid quarter-century delving into the controversies of evolutionary theory (and reading just about everything else that Dawkins has written, along with untold numbers of books and essays by other evolutionists and their critics), I wondered if it would still strike me as brilliant, or would I now see flaws, over-simplifications, solecisms that escaped my naive reading? Having climbed the ladder, would I now want to discard it?

So I took a copy along on a two-week trip, in June 2005, to the Galápagos, for a cruise arranged by the historian of science, Frank Sulloway (whose work had upset the traditional myth of Darwin's *eureka* moment while on the Beagle), followed by the World Summit on Evolution organized by the Universidad San Francisco de Quito, held on San Cristóbal Island. I would be spending my days in conversation with some of the world's best evolutionary biologists, and I'd be rereading *The Selfish Gene* while sailing from island to island, in Darwin's footsteps. I didn't take my heavily underlined and annotated 1976 edition, but my almost equally heavily underlined and annotated 1989 edition, with all the endnotes and the two additional chapters. What follows are my reflections on this rereading, most of them composed on my laptop in the salon of Sagitta, a gracious three-masted schooner, while riding at anchor in one or another of the Beagle's stopping places in the Galápagos. With all the intellectual and perceptual competition, this would be a stern test for any book.

What struck me most was that my deep appreciation of Dawkins has been strengthened, not diminished, by the interven-ing years of evolutionary adventures. I found myself in rousing agreement with Dawkins' opening passage: 'We are survival machines—robot vehicles blindly programmed to preserve the selfish molecules known as genes. This is a truth which still fills me with astonishment.'[6] As he went on to say in the preface to the 1989 edition,[7] the theorists whose work he celebrates in the book had clearly articulated this truth, 'But I found their expres-sions of it too laconic, not full throated enough'. What Dawkins saw was that Darwin's scientific revolution was also a philosophical revolution: 'Zoology is still a minority subject in universities, and even those who choose to study it often make their decision without appreciating its profound philosophical significance.'[8] Indeed. Darwin's dangerous idea amounts to nothing less than a reframing of our fundamental vision of ourselves and our place in the universe. Stephen Jay Gould once branded us both as 'Darwinian fundamentalists',[9] and in spite of the negative connotations Gould intended to convey by that epithet, there is a sense in which he was right. *This* is the fundamental truth of Darwinism, and, as I have tried to show in my own work, there are no stable intermediate positions; either you shun Darwinian evolution altogether and cling to an Aristotelian or Abrahamic vision of God as Prime Mover and Creator, or you turn that traditional universe upside-down and accept that mind, meaning, and purpose are the fairly recent *effects* of the churning mechanistic mill of mindless Darwinian algorithms, not their *cause*. Design is generated originally by bottom-up processes, and all the top-down processes of research and development that we know so well (human authorship and exploration, invention, problem-solving, and creation) are themselves the evolved fruits of these bottom-up processes at many levels and scales, including Darwinian algorithmic processes within individual brains. All the attempts at compromise, at making exemptions for one cherished treasure or another by hanging it on a skyhook, are doomed to incoherence.

I was invited by Dawkins to write an Afterword for the new edition (1999) of *The Extended Phenotype*, and I opened that brief essay much as I have opened this one, by applauding both the philosophical methods and the very considerable philosophical substance of that work.

Why is a philosopher writing an Afterword for this book? Is *The Extended Phenotype* science or philosophy? It is both; it is science, certainly, but it is also what philosophy should be, and only intermittently is: a scrupulously reasoned argument that opens our eyes to a new perspective, clarifying what had been murky and ill-understood, and *giving us a new way of thinking* about topics we thought we already understood.[10]

This and other published instances of commendation earned me the epithet 'Dawkins' lapdog' from Gould. Huxley was happy to call himself Darwin's bulldog, and Dawkins has shown that he can be his own bulldog, so I'd be happy to be known as a hard-working sled dog on the same team as Dawkins; but enough about dogfights. Our agreement is deep and detailed, and I give him credit for some of the issues we chime together on, but on some others I got there on my own. What matters is not who got what first, but that our convergent versions support each other, gaining strength from the different paths by which we arrived at them.

Most centrally, consider our *mentalistic behaviorism*. If you think that's a contradiction in terms, you've missed the boat. To see why, you have to appreciate an earlier scientist/philosopher friendship, between B. F. Skinner and W. V. O. Quine. Fred Skinner's brand of behaviorism was a philosophically driven methodology: according to him, mind-talk in all its varieties was dualistic and *mysterian* (to speak anachronistically, using Owen Flanagan's useful term). Science was to be materialistic, and mechanistic, and hence should abjure all use of mentalistic idioms.[11] Never speak of an animal as *knowing* or *wanting* or *believing* or *expecting*. Speak instead of an animal's dispositions to behave (where the behavior was to be described in scrupulously mechanistic terms—no *seeking* behavior or *investigating* behavior, for instance. His Harvard colleague and friend Van Quine appreciated Skinner's impatience with unsupported mind-talk, and sharpened that puritan ethic by analyzing mind-talk as logically pathological: the *intentional idioms* of mentalistic discourse exhibited the awkward feature of 'referential opacity'. (In a sentence such as 'Tom believes that Tully denounced Catiline', you can't 'substitute equals for equals' by putting in 'Cicero' for 'Tully'—two expressions referring to the same individual—and be sure that you will preserve the truth, as you can always do in a normal, referentially transparent sentence.) In a famous phrase, Quine joined forces with Skinner and threw down the gauntlet:

One may accept the Brentano thesis [of the irreducibility of intentional idioms] either as showing the indispensability of intentional idioms and the importance of an autonomous science of intention, or as showing the baselessness of intentional idioms and the emptiness of a science of intention. My attitude, unlike Brentano's, is the second.[12]

This abstemious brand of behaviorism, apparently the straight-forward extension of what we might call standard scientific positivism, had widespread influence in the second half of the twentieth century, but meanwhile another sort of behaviorism was developing, which helped itself blithely to a few well-chosen intentional idioms—mainly *expect* and *prefer* (or, equivalently in philosopher-speak, *believe* and *desire*)—and built up impressive edifices of theory and practice, in decision theory and game theory, in economics and computer science and cognitive science, and farther afield. This was close kin to the 'logical behaviorism' informally explored by Gilbert Ryle in *The Concept of Mind*.[13] Although it is unclear whether Ryle had much influence outside philosophy, his coupling of staunch anti-Cartesianism (the notorious 'ghost in the machine') with an insouciant disregard of the Skinnerian strictures opened the conceptual floodgates. The key idea, as he saw, was that mentalistic terms were a convenient way of speaking of *dispositions* to behave, and more particularly behavioral *competences* or *abilities*. When one spoke of somebody's knowledge or expectation, goals or preferences, one was alluding not to some spooky, metaphysically private inner goings-on, but to a pattern of (mainly intelligent) *action* that could be expected from this agent. Although Ryle concentrated on human minds, the extension of this perspective to other natural phenomena, both 'higher' and 'lower', was more or less guaranteed by the substrate-neutrality or abstractness of any such dispositional analysis: handsome is as handsome *does*, and anything that can behave *as if striving* for this and that while guided by what it '*knows*' is an appropriate subject for such an analysis.

So, just as a person could be confirmed to be *vain* by observing how she tended to act when in the presence of mirrors or potential admirers (without our having to look into her soul for some imagined vanity-nugget), a gene could be *selfish* without our having to impute any consciousness or 'qualia' or other dubious mental furniture to it: 'It is important to realize that the above definitions of altruism and selfishness are *behavioural*, not subjective. I am not concerned here with the psychology of motives.'[14] Dawkins' brilliant application of mentalistic behaviorism—what I call the *intentional stance*—to evolutionary biology was, like my own coinage, an articulation of ideas that were already proving themselves in the work of many other theorists. We are both clarifiers and unifiers of practices and attitudes pioneered by others, and we share a pantheon: Alan Turing and John von Neumann on the one hand, and Bill Hamilton, John Maynard Smith, George Williams, and Bob Trivers on the other. We see computer science and evolutionary theory fitting together in excellent harmony; it's algorithms all the way down.

Dawkins and I have both had to defend our perspective against those who cannot fathom—or abide—this *strategic* approach to such deep matters. Mary Midgley[15] was incredulous—how on earth could a gene be *selfish*?—while John Searle[16, 17] was equally scornful—anybody who says a thermostat has a *belief* about the temperature must be crazy! Mere *as if* intentionality or *derived* intentionality could never explain our *real*, *original* intentionality. But the appropriate response to this incredulity is the same response that any biologist should make to a similar challenge about where to 'draw the line' between the living and the non-living: 'Should we call the original replicator molecules "living"? Who cares?'.[18] One of the central lessons of Darwinian thinking is that essentialism must be abandoned: the imagined 'essence of life' has to be approached by one imaginable chain or another of simple agents or agencies stretching from the clearly non-living to the clearly living, and only a lexicographical *decision* is going to 'draw the line'. There are better and worse joints at which to carve nature, but they are better only in that they make life easier for the theorist. As we climb the scale from utterly mindless (but selfish) genes through almost equally mindless (but still striving) macro-molecular mousetraps to ingeniously designed (but still clueless) fledgling cuckoos to clever apes (and robots) to wonderful, mind-ful *us*, if anyone asks us the question 'But which of these inten-tional systems have *real* minds?' the answer is: 'Who cares?' There couldn't be 'real' minds and 'real' selfishness without billions of years of hemi-semi-demi-pseudo-proto-quasi-minds and mere 'as if' selfishness to drive the research and development process that has eventually yielded our minds. Now we can look back: our paradigmatic minds and purposes can be used as our model—sometimes literal, sometimes metaphorical, sometimes semi-literal/semi-metaphorical—for the processes that populated the ancestral phenomena. The same virtual regress can be played out synchronically by treating a human mind (for instance) as a 'soci-ety of minds'[19, 20]—lesser intentional agencies competing and cooperating in the ongoing task of maintaining a *soul* to govern a body, and those clever striving agencies are made up of simpler agencies in turn, and so on until we get to functionaries so simple and mindless that they can be replaced by a machine.

'Yes, we have a soul; but it's made of lots of tiny robots!'[21]—and it was designed by a Blind Watchmaker. Championing such vivid oxymorons is not just a rhetorical habit that Dawkins and I share; it is a deliberately designed assault on the default presumption of the pre-Darwinian world: the trickle-down vision in which all Design must come from a greater, higher Mind, instead of bubbling up from mindless, motiveless mechanisms.

One of the insights I gained from rereading *The Selfish Gene* in the context of the World Summit on Evolution was that it is not just the Midgleys and Searles who are uncomfortable with Dawkins' anthropomorphizing of genes; there are eminent evo-lutionary scientists who still yearn for a biological version of stripped-down Skinnerian/Quinian behaviorism. They may not know just why they are unwilling to speak, with Dawkins, about what the Blind Watchmaker has *discovered* over and over again (in convergent evolution), and they fully appreciate the aptness of

Orgel's Second Rule (evolution is cleverer than you are), but they feel a little guilty indulging in such talk, even in the squeaky-clean contexts of evolutionary game theory. (On several occasions in discussion with such self-styled hardheads I have been put in mind of the philosopher Sydney Morgenbesser's reaction to such purit-anical overkill: 'Let me see if I've got this straight, Prof. Skinner: you're saying it's a mistake to anthropomorphize *humans*?' The Skinnerians did seem to think that in order to be properly scien-tific they had to pretend that people were stupid, since after all, at bottom, people are made of nothing but atoms and atoms are stupid. And some evolutionists still seem to think that they have to refrain from using the oft-proven fact that natural selection can be relied upon to *find* the best move in the design problems set for it.[22] Yes, at bottom, evolution is a mindless, purposeless, mechanistic process, but at higher levels of analysis, it can be seen to be teem-ing with agent-like entities engaged in competitions, exploring possibilities, solving problems, discovering designs.)

Dawkins, unlike many scientists—and most philosophers—is comfortable with definitions that lack the hard edges of necessary and sufficient conditions. Even the central concept of a gene, he claims, can get by on 'a kind of fading-out definition, like the definition of 'big' or 'old'.[23] Is this really acceptable? Philosophers have a way of starting off down a promising path and then stop-ping after the first few steps and spending the rest of their time and energy worrying about some problem of definition or an assumption that they might better just grit their teeth and *make*! Dawkins goes on: 'A gene is not indivisible, but it is seldom div-ided.'[24] That is what makes it a gene, in fact: its salience over longish periods of time. (That is, it is the salience over time of a particular undivided but still varying sequence that makes it the case that *there is something there worth reidentifying and nam-ing*.) 'The gene is defined as a piece of chromosome which is sufficiently short for it to last, potentially, *for long enough* for it to function as a significant unit of natural selection.'[25] But notice that it is not a particular hunk of DNA Dawkins is talking about: philosophers would say he's talking about a *type* not a *token*.

(Two tokens of the word type 'talking' occur in the previous sentence. When I say that 'talking' is a two-syllable word, I'm talking about the type, not any particular token.) Dawkins puts the point this way: 'What I am doing is emphasizing the potential near-immortality of a gene, in the form of copies, as its defining property.'[26] So genes are like words, or like novels or plays, or melodies. A particular play, such as *Romeo and Juliet*, exists in many tokens, on stages and in books, on videotapes and DVDs. A particular gene also exists in many tokens, in trillions of cells. This is Dawkins' way of making George Williams' point that the gene is the *information* carried in the base pairs, not the base pairs themselves, which are like the trails of ink (or acoustic waves or laser-readable DVD pits). What counts as the gene is not just any canonical reading of the text but also all the mutant misreadings that are clearly misreadings *of this text*.

Dawkins is equally pragmatic in his treatment of animal signals. Some theorists have wondered if we can ever say exactly what any animal signal *really* means, or what any animal is *really* thinking, but Dawkins wisely avoids committing himself to this hysterical realism: 'If we wish to (it is not really necessary), we can regard signals such as the cheep call as having a meaning, or as carrying information . . .'.[27] Alarm calls 'could be said to carry information'.[28] While I approve of this reluctance to be drawn into definitional battles, such reluctance can be carried too far, leading one to overlook or underestimate important differences. For instance, in his brief discussion of a lecture he attended by Beatrice and Allen Gardner, trainers and keepers of the famous signing chimpanzee Washoe,[29] he mentions, disparagingly, the philosophers at the lecture who 'were very much exercised by the question of whether Washoe could tell a lie'. He suspects that the Gardners thought there were more interesting things to talk about, and says he agreed with them. It was not important, he suggests, to inquire into whether Washoe could tell a deliberate lie, knowingly and consciously intending to deceive. What was interesting, he suggests, is just creating an 'effect functionally equivalent to deception'.

Angler fish wait patiently on the bottom of the sea, blending in with the background. The only conspicuous part is a wriggling worm-like piece of flesh on the end of a long 'fishing rod', projecting from the top of the head. When a small prey fish comes near, the angler will dance its worm-like bait in front of the little fish, and lure it down to the region of the angler's own concealed mouth. Suddenly it opens its jaws, and the little fish is sucked in and eaten. The angler is telling a lie, exploiting the little fish's tendency to approach wriggling worm-like things. He is saying 'Here is a worm', and any little fish who 'believes' the lie is quickly eaten.[30]

This is true, and a fine use of the intentional stance, but it is also true that deliberate lies are on a different plane from the functional deception of angler fish. There are many intermediate cases of quasi-knowing deception in animals—the distraction displays of such low-nesting birds as piping plovers are a well-studied instance[31]—as well as a bounty of tempting anecdotes about the 'Machiavellian' intelligence of primates.[32] In fact the question of whether Washoe could tell a deliberate lie is a deeply interesting theoretical question, investigated at length with another chimpanzee, Sarah, by David Premack and his colleagues, and leading to some intermittently fruitful and important research on both animals and children, the ill-named 'theory of mind' controversy.[33] The transition from mindless deceit to mindful deceit is a good manifestation of a major transition in evolution—not a metaphysical or cosmic distinction, an unbridgeable chasm, but a passage, with intermediate transitional cases of deceit that may not be so mindless. Once that transition has been clearly accomplished, it opens up a whole new world of deceit (and other sophisticated behavior). Dawkins recognized this himself in his commentary in *Behavioral and Brain Sciences* on 'Intentional Systems in Cognitive Ethology',[34] so this is not a point of ongoing disagreement.

In fact, of course, Dawkins' insight into the role of cultural evolution in designing the minds of one species of primates, *Homo sapiens*, has been a major influence on my own work. The concept of a meme, a replicating unit of cultural evolution that can move from brain to brain, redesigning the brain a little to make it a better outpost for itself and other memes, opens up ways of thinking about psychological phenomena—both cognitive and emotional—that were inaccessible to earlier theorists puzzling about the problems of consciousness. Now that we have the idea, it even seems obvious, in retrospect, that most of the huge difference between our minds and the minds of chimpanzees is not due directly to the genetically controlled differences in neuroanatomy but to the vast differences in *virtual* architecture made possible by those minor differences in the underlying neural hardware. By becoming adapted to the transmission and rehearsal (internal replication) of a cornucopia of pre-designed cultural thinking tools, our brains became open-minded in a way that is apparently unavailable to chimpanzee brains no matter how intensively their cultural environment is enriched.

At this time, the contributions of the concept of a meme are still largely conceptual—or philosophical. The search for testable hypotheses of memetics is still in its infancy, but there are more than a few applications of the underlying insights to theoretical problems in philosophy, cognitive science, and more recently, the nature of ethics and religion.[35] For instance, I recommend Balkin's *Cultural Software*,[36] and my own book on religion as a natural phenomenon, *Breaking the Spell*.[37] The creation of a new scientific concept is like speciation: you can't identify a successful instance at the moment of birth. Time will tell whether, in another century, Dawkins' chapter on memes will be retrospectively crowned as the birth of an important scientific lineage of work. I am betting on it, but what about my claim that the book is excellent philosophy in any case? A psychologist colleague, on reading a draft of this essay, asked if *The Selfish Gene* is considered required reading in any philosophy graduate program. Certainly specialists in the philosophy of science or philosophy of biology would be expected to have read it, but what about students of epistemology or philosophy of mind or language? We philosophers are a somewhat conservative lot, loath to grant that anybody but a professional philosopher could write something worthy of entry into the canon. If you put *The Selfish Gene* on the *required* reading list, just which 'classic' would you bump from the list to make room for it? I have seen enough philosophy students enthusiastically tell me how they were transformed by reading the book to judge that it pulls its weight and then some, so yes, I put Dawkins' book alongside classics by such non-philosophers as Turing[38] and Kuhn[39] as essential thinking tools for any student of philosophy. In addition to everything else they will learn from it, they will discover that it is actually possible to write arguments that are both rigorous and a joy to read. That discovery, if enough philosophers took it to heart, could transform our discipline.

ENDNOTES

1 Richard Dawkins, *The Selfish Gene* (Oxford: Oxford University Press, 2nd edn., 1989), 322.

2 Douglas R. Hofstadter and D. Dennett, *The Mind's I: Fantasies and Reflections on Self and Soul* (New York: Basic Books, 1981).

3 Daniel Dennett, *Content and Consciousness* (London: Routledge & Kegan Paul, 1969).

4 Daniel Dennett, 'Intentional Systems', *Journal of Philosophy*, 68 (1971): 87–106.

5 Daniel Dennett, 'Why the Law of Effect Will not Go Away', *Journal for the Theory of Social Behaviour*, 5 (1975): 169–187.

6 Dawkins, *The Selfish Gene* (1st edn., 1976), v.

7 Dawkins, *The Selfish Gene* (2nd edn., 1989), ix.

8 Dawkins, *The Selfish Gene* (2nd edn., 1989), 1.

9 Stephen Jay Gould, 'Darwinian Fundamentalism', *New York Review of Books* (12 June 1997), 34–37 and 'Evolution: The Pleasures of Pluralism', *New York Review of Books* (26 June 1997), 47–52.

10 Daniel C. Dennett, 'Afterword', in Richard Dawkins, *The Extended Phenotype* (Oxford: Oxford University Press, 1999).

11 For a detailed analysis of Skinner's position, see Daniel C. Dennett, 'Skinner Skinned', in *Brainstorms* (Cambridge, MA: MIT Press, 1978).

12 Willard Van Orman Quine, *Word and Object* (Cambridge, MA: MIT Press, 1960), 227. Quine acknowledged the 'practical indispensability'

in daily life of the intentional idioms of belief and desire but disparaged such talk as an 'essentially dramatic idiom' rather than something from which real science could be made in any straightforward way (*Word and Object*, page 217), but later in his career he came to appreciate that this dramatic idiom might be harnessed into a predictive science. My own very Quinian analysis of the philosophical problems of referential opacity and their dissolution can be found in two essays: Daniel C. Dennett, 'Beyond Belief' and 'Mid-term Examination: Compare and Contrast', in *The Intentional Stance* (Cambridge, MA: MIT Press/Bradford, 1978).

13 Gilbert Ryle, *The Concept of Mind* (London: Hutchinson, 1949).

14 Dawkins, *The Selfish Gene* (2nd edn., 1989), 4.

15 Mary Midgley, 'Gene-juggling', *Philosophy*, 54 (1979): 439–458.

16 John Searle, *Minds, Brains and Science* (Cambridge, MA: Harvard University Press 1985).

17 John Searle, *The Rediscovery of the Mind* (Cambridge, MA: MIT Press, 1992).

18 Dawkins, *The Selfish Gene* (2nd edn., 1989), 18.

19 Marvin Minsky, *Society of Mind* (New York: Simon & Schuster, 1985).

20 George Ainslie, *Breakdown of Will* (Cambridge: Cambridge University Press, 2001).

21 'Si abbiamo un anima. Ma è fatta di tanti piccolo robot.' Giulio Giorelli, quoted in Daniel C. Dennett, *Freedom Evolves* (London: Penguin, 2003), 1.

22 Richard Lewontin has often insisted that it is a mistake to assume that lineages are posed problems by the environments they encounter, and since a wealth of adaptationist success stories belies this claim, it is tempting to conclude that Lewontin views all these results as ill-gotten gains. For a discussion, see Richard Lewontin, 'Elementary Errors about Evolution', *Behavioral and Brain Sciences*, 6 (1983): 367–368 and Daniel C. Dennett, (1983), my response to his response.

23 Dawkins, *The Selfish Gene* (2nd edn., 1989), 32.

24 Dawkins, *The Selfish Gene* (1989), 33–34.

25 Dawkins, *The Selfish Gene* (1989), 35–36.

26 Dawkins, *The Selfish Gene* (1989), 35.

27 Dawkins, *The Selfish Gene* (1989), 63.

28 Dawkins, *The Selfish Gene* (1989), 64.

29 Dawkins, *The Selfish Gene* (1989), 64.

30 Dawkins, *The Selfish Gene* (1989), 64–65.

31 C. Ristau, 'Aspects of the cognitive ethology of an injury-feigning bird, the piping plover', in C. Ristau (ed.), *Cognitive Ethology: The minds of other animals* (Hillsdale: LEA, 1991).

32 Andrew Whiten and R. Byrne (eds.), *Machiavellian Intelligence* (Oxford: Oxford University Press, 1988) and *Machiavellian Intelligence, II: Extensions and Evaluations* (Cambridge: Cambridge University Press, 1997).

33 Daniel C. Dennett, 'Beliefs about Beliefs' (commentary on Premack, Woodruff, et al.), *Behavioral and Brain Sciences*, 1 (1978): 568–570.

34 Richard Dawkins, 'Adaptationism was always predictive and needed no defense', *Behavioral and Brain Sciences*, 6 (1983): 360–361. Commentary on Daniel C. Dennett, 'Intentional Systems in Cognitive Ethology: The "Panglossian Paradigm" Defended', *Behavioral and Brain Sciences*, 6 (1983): 343–390.

35 For a survey of the status and progress of memetics, see Daniel C. Dennett, 'The New Replicators', in Mark Page (ed.), *The Encyclopedia of Evolution* (Oxford: Oxford University Press, 2002), vol. 1, E83–E92.

36 J. M. Balkin, *Cultural Software: A Theory of Ideology* (New Haven: Yale University Press, 1998).

37 Daniel C. Dennett, *Breaking the Spell* (London: Penguin, 2006).

38 A. Turing, 'Computing Machinery and Intelligence', *Mind*, 59 (1950): 433–460.

39 Thomas Kuhn, *The Structure of Scientific Revolutions* (Chicago: University of Chicago Press, 1962).

THE INVENTION OF AN
ALGORITHMIC BIOLOGY

Seth Bullock

BIOLOGY and computing might not seem the most comfortable of bedfellows. It is easy to imagine nature and technology clashing as the green-welly brigade rub up awkwardly against the back-room boffins. But collaboration between the two fields has exploded in recent years, driven primarily by massive investment in the emerging field of bioinformatics charged with mapping the human genome. New algorithms and computational infrastructures have enabled research groups to collaborate effectively on a worldwide scale in building huge, exponentially growing genomic databases, to 'mine' these mountains of data for useful information, and to construct and manipulate innovative computational models of the genes and proteins that have been identified. This recent burst of high-profile activity might suggest that computer scientists have only just begun to work on biological questions, but activity at this particular disciplinary interface is by no means new. In fact, it has an extremely long history involving the most famous early pioneers of computing, cybernetics, and artificial intelligence.

In the 1950s, Alan Turing, the 'father of artificial intelligence' and a man fundamentally associated with codes, logic, chess, and other mechanico-mathematical arcana, developed influential models of biological morphogenesis:[1] the processes involved in the development of biological patterns as an organism grows from a single cell. He was particularly interested in accounting for the tendency of spiral patterns in many plant structures to obey the Fibonacci sequence (e.g. if you count the number of whirls running clockwise on a pine cone and the number running anticlockwise, the two numbers will be consecutive terms in Fibonacci's famous sequence of integers: 0, 1, 1, 2, 3, 5, 8, 12, . . .). At the same time, John von Neumann, one of history's great polymaths and the man responsible for game theory and the architecture of the modern computer among many other things typically considered to lie far from the muddy field of biology, worked on the problem of self-replication:[2] over evolutionary time, simple life-forms have given rise to more complicated creatures, but how, von Neumann asked, could a machine (like a dog or an amoeba or a robot) make a more complex version of itself? The answer that he arrived at predicted the essential distinction between DNA (instructions) and transcriptase (machinery that follows instructions) several years before Crick and Watson's discovery.

Surprisingly, though, the very first example of activity fusing computing and biology is over a century older than the work of Turing and von Neumann, predating even Darwin's *Origin of Species*. It is due to Charles Babbage, designer of the Difference Engine, the first automatic calculating machine and the progenitor of the modern computer. As early as 1837, Babbage reported using this machine to help him demonstrate that inexplicably abrupt changes in the geological record need not be taken to be the work of God (a hot topic of the day). He showed that his completely deterministic (clockwork) machine could generate analogous surprising behavior ('miracles') without any external interference from the programmer. He invited his contemporaries (including Darwin) to observe the machine generating a sequence of numbers (1, 2, 3, 4, . . .) and asked them to state the rule or law that the machine was obeying. At some predetermined point the engine would 'disobey' this law, automatically beginning to generate some alternative stream of numbers (the Fibonacci sequence, perhaps), and surprising the onlookers, who were forced to admit that apparently mysterious and abrupt changes observed in nature need not demand explanation in terms of divine intervention.[3]

It is clear, then, that computing and biology have communicated from almost the first possible moment, and have been finding new and productive ways to interact ever since. And it is firmly within this tradition that some of Richard Dawkins' most interesting work can be located. Indeed, there are two senses in which this is true. First, and most straightforwardly, Dawkins has had a significant involvement in the development of *bio-inspired algorithms*, specifically within the field of evolutionary computation, where computer programs solve problems in a manner inspired by biological evolution. In 1986, in *The Blind Watchmaker*, Dawkins introduced an algorithm of the same name. This computer program requires a user repeatedly to select one of nine bilaterally symmetrical line drawings, or 'biomorphs' (see Fig. 1). After each selection, nine new variants of the chosen biomorph are randomly generated and presented. Over time, the line drawings 'evolve' to reflect the taste of the user, who is effectively breeding biomorphs by exerting selection pressure on a population of forms that are competing with one another for the chance to 'reproduce'.

A year later, Dawkins presented his biomorphs at an 'Interdisciplinary Workshop on the Synthesis and Simulation of Living

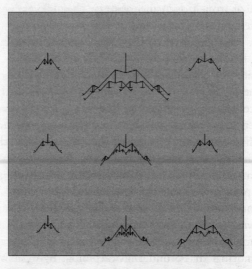

1. Biomorphs 'evolved' using the *Blind Watchmaker* program.

Systems' held at Los Alamos National Laboratory in New Mexico. The meeting brought together a disparate group of researchers from computing, mathematics, physics, biology, neuroscience, and even economics to talk about a set of topics that have come to be known collectively as Artificial Life.[4] What is life? Can it be synthesized *in silico*? What can we learn about life in the attempt? Dawkins' involvement at the outset of artificial life (along with that of other biologists such as Elliot Sober and John Maynard Smith) leant the field some credibility, but his contribution[5] to the first conference is also notable in its own right. In it, he presented the *Blind Watchmaker* program as a tool with which to explore the notion of *evolvability*—the tendency of a population to tolerate and eventually profit from small changes (mutations). This property remains poorly understood. While biological progeny are not identical to their parents or their siblings, they typically remain viable organisms. By contrast, introducing a few random mutations into a computer program or a hospital's working procedures is likely to prove catastrophic. Moreover, the mutations suffered by biological organisms are not just neutralized, corrected, or ironed out, since enough useful variation amongst relatives remains to fuel natural selection. This balance between robustness and sensitivity, between staying the same and changing, has yet to be understood and exploited in evolutionary computation or other relevant fields—amongst other things, a full understanding of it would revolutionize our ability to manage evolving complex systems such as hospitals, cities, economies, and so on. Dawkins' paper represents an early attempt to address some of these issues.

Dawkins' program itself is unusual in that, unlike standard evolutionary algorithms, it demands that the user *manually* exert selection pressure on an artificial evolving population, choosing which 'biomorphs' get to reproduce. This approach has inspired a whole *oeuvre* of 'aesthetic evolutionary algorithms' in which artists produce their art in *partnership* with an artificial evolutionary process, moving far beyond Dawkins' stick figures, to generate much more complex pieces[6] (see Fig. 2). Our commonsense notion

2. Evolved artwork. © Karl Sims, used by kind permission.

of artistic creativity combines both a *generative* aspect (actually making, altering, improving the artefact) with a *selective* aspect (choosing whether the alteration makes the artefact better, or complete). By contrast, Dawkins' evolutionary approach cedes responsibility for generation to the computer which randomly (rather than purposively) perturbs the currently selected individual. The artist reserves only the right to sift these perturbed forms and select which of them are to be (mis)copied into the next generation.

As such, in addition to serving as a tool with which to introduce adaptation by natural selection to a general audience, the program raises a number of interesting questions concerning progress, purpose, and creativity in art and nature. Is the user of such a computer program really an artist, and if so what is the status of the program's writer? Can pointing and clicking one's way through a (potentially infinite) genetic space of 'predefined' forms be somehow equivalent to painting or drawing? In fact, the practice resembles the (non-artistic?) selective breeding of plants, livestock, or domestic animals, but simultaneously resonates with some experimental art in which the artist's volition is similarly attenuated (e.g. Jackson Pollock's action painting, which coupled spontaneous 'random' splashing and dripping with careful subsequent editing, cropping, or outright rejection).[7]

While Dawkins' simple computer program was the first example of a commercially released piece of artificial life software, its potency is better evidenced by the number of times it has been recoded and extended by those who have read about it. The internet is home to a veritable cottage industry of biomorph breeding, and many programmers (including my teenage self) must have written their variants of the *Blind Watchmaker* before the internet allowed them to be widely disseminated. There is something compelling in the combination of simplicity, scope, and visual impact that captures the imagination of these programmers, and comes to influence the way that they think about evolutionary processes and algorithms. This is the second sense in which Dawkins' work lies at the boundary between computing and biology—the pedagogical use of specific algorithms and algorithmic thinking and talking to understand and explain evolutionary biological processes: what might be termed algorithmic biology.[8]

An algorithm is a set of step-by-step instructions, like a cake recipe or travel directions. As such, our tacit understanding is that they are useful, but inert and straightforward. Dawkins employs an algorithmic device explicitly when he describes, in *The Blind Watchmaker*, how a particular string of symbols (the sentence: 'Methinks it is like a weasel') might arise via reproduction, mutation, and selection in a population of initially random symbol strings.[9] By repeatedly applying the same sequence of actions, the appearance of deliberate design is achieved despite the randomness inherent to the process and the vast number of possible sentences (roughly 27^{28} if we don't care about upper case or punctuation). Like his biomorphs program, this algorithm is a powerful rhetorical device because it mechanizes and thereby demystifies natural selection (at the expense, perhaps, of muddying the waters concerning the nature of biological selection pressures, which are neither aesthetic nor aiming at a prearranged target).

There are, of course, alternative ways of conveying the central tenets of natural selection: drawing parallels with selective breeding of pigeons or flowers; conveying the impact of finite resources on heritable variation; explaining the implications of the second law of thermodynamics for copying processes. Dawkins makes use of many of these, but algorithmic devices are special. One of their key features is that they are *multiply realizable*. This just means that the same algorithm can be carried out by many different machines. You or I could follow the same set of directions and your computer or mine could execute the *Blind Watchmaker* program (if the languages in which the algorithms are written are appropriate). Algorithms abstract away from nitty-gritty implementation details (just where is my cake tin? how exactly do I 'jump on the No. 1 bus'?), casting a process at a level that rises somewhat above particular instances of execution, without resorting to mathematical or logical formalisms that have limited currency.

Crucially, when taking an algorithmic approach to natural selection, rather than writing in terms of, for example, competition for scarce resources (fighting, fleeing, feeding, sex), the evolutionary process is free to dissociate from the 'four Fs', thereby becoming readily applicable to a wider range of non-genetic (quasi-)evolutionary systems. Most famously, and much earlier, in *The Selfish Gene* Dawkins was able to reapply the abstracted principles of natural selection within the realm of ideas, conjuring the *meme* as an ideational equivalent of the biological gene.[10] Since then, there has been a significant proliferation of (quasi-) evolutionary approaches to a range of non-genetic systems: evolutionary linguistics, economics, psychology, and even cosmology, as well as evolutionary computation and art. In most cases, the success or failure of these enterprises cannot yet be judged, but their very existence is testament to the expanding power of uprooted evolutionary biological concepts and, in particular, the biological algorithm at the heart of evolution by natural selection.

The pioneers name-checked at the outset of this paper suggest that, historically, most significant workers at the computing–biology interface have tended to be mathematicians or computer scientists who are interested in biological questions. Dawkins bucks this trend somewhat, in that he is a biologist, and one who has not been particularly interested in computational questions. Rather, he is interested in *using* computers, not just as tools with which to write or calculate, but primarily as tools with which to think.

ENDNOTES

1 A. M. Turing, 'The chemical basis of morphogenesis', *Philosophical Transactions of the Royal Society of London, Series B,* 237 (1952): 37–72.

2 J. von Neumann and A. W. Burks, *Theory of Self-Reproducing Automata* (Urbana, IL: University of Illinois Press, 1966).

3 Babbage's automated 'miracles' did, however, suggest an image of God as an omnipotent engineer, having no need for divine intervention, but instead relying upon 'pre-programmed' miraculous events. At the time, it remained an act of faith to believe that this kind of pre-programming could, unlike fundamentally inexplicable acts of divine intervention, be understood through regular science (in the same way that careful examination of Babbage's engine would have revealed the laws that governed its surprising behavior). For more discussion, see S. Bullock, 'Charles Babbage and the emergence of automated reason', in P. Husbands, O. Holland, and M. Wheeler (eds.), *The History of the Mechanization of Mind* (Cambridge, MA: MIT Press, forthcoming).

4 C. G. Langton (ed.), *Artificial Life: An Overview* (Cambridge, MA: MIT Press, 1995).

5 R. Dawkins, 'The evolution of evolvability', in C. G. Langton (ed.), *Artificial Life: The Proceedings of an Interdisciplinary Workshop on the Synthesis and Simulation of Living Systems* (Redwood City, CA: Addison-Wesley, 1989).

6 K. Sims, 'Artificial evolution for computer graphics', *Computer Graphics*, 25/4 (1991): 319–332.

7 A. Dorin, 'Aesthetic fitness and artificial evolution for the selection of imagery from the mythical infinite library', in J. Kelemen and P. Sosík (eds.), *Proceedings of the Sixth European Conference on Artificial Life* (Berlin: Springer, 2001), 659–668.

8 I owe the term to Richard A. Watson, although my usage may not be equivalent to his.

9 R. Dawkins, *The Blind Watchmaker* (New York: W. W. Norton, 1986).

10 R. Dawkins, *The Selfish Gene* (Oxford: Oxford University Press, 1976).

SELFISH GENES AND INFORMATION FLOW

David Deutsch

ALONG with countless other people, I had been labouring under some significant misconceptions before I relearned the theory of evolution from Richard Dawkins' book *The Selfish Gene*.

Many of my own interests have concerned *information flow*—how information gets from one place to another and how it changes from one form into another. I did not always think of it in these terms at the time, but, for example, one of the fields in which I have worked is the 'parallel-universes' interpretation of quantum theory. It says that the universe that we see around us is part of a much larger structure, the 'multiverse', which contains many such universes, some like ours, some different. And I became convinced of this theory essentially by regarding the world as a system of information flow: if one analyses this flow under quantum theory, it turns out to consist of vast numbers of sub-flows that are nearly autonomous. That is to say, the behaviour of each of them over time depends almost entirely on its own state, and very little on the state of the others. Moreover, the information in each of these sub-flows behaves very like that which would define the universe of classical physics. Because they are not perfectly parallel—they do affect each other through quantum interference effects—one must regard all of them as equally real, and so the multiverse conclusion is inescapable.

Another example is that I learned, by reading Karl Popper, how knowledge (which is one kind of information) is created with the help of evidence (another kind of information), through conjectures and experimental tests. I learned that this process cannot possibly take the form of extrapolation ('induction') of the evidence into knowledge.

Lamarckism (evolution through the inheritance of character-istics acquired when organisms interact with their environments) is the same mistake in the theory of evolution as 'induction' is in the philosophy of science. Both involve what Popper called 'instruc-tion from the environment': observations supposedly instruct a scientist on what theory would explain them, and an organism's experiences supposedly instruct it on what changes would adapt to them. In both cases, *how* one form of information gets itself translated into the other is not addressed: it just does. Both of these mistaken theories get the direction of information flow wrong: they assume that the new knowledge or adaptation starts outside the brain or organism and flows in, while in reality it starts inside, as a conjecture or mutation, and the environment has no effect until after that information is already in existence. The only role of the environment is to choose between different conjectures or mutations that are otherwise viable. Analogously, the traditional ways of trying to cling to the single-universe con-ception of reality in quantum physics depend on explicitly eschewing any description of how information flows in that uni-verse: how the observed results of experiments come about as a result of certain initial conditions. Again, the idea is that they just do.

People are divided more by the different problems that they consider worth addressing than by the different theories that they advocate. Dawkins is closer, philosophically, to William Paley, the pre-Darwinian creationist and cogent advocate of the Argument from Design, than he is to Lamarck, the pre-Darwinian evolutionist. That is because Paley and Dawkins (and of course Darwin) all understood what the problem is: to explain how the knowledge (or 'design') in biological adaptations could possibly come into existence spontaneously. Paley thought he could prove that it could not; but that is a relatively unimportant detail. If, ahistorically, we express Lamarck's conception of the problem in terms of information flow, then he was asking how the knowledge of how to prosper in an environment gets from that environment into the organisms that live there. Since information does not and cannot flow in that manner, no way of answering that question could ever have explained anything.

Therefore, merely rejecting wrong *answers* such as creationism and Lamarckism is not enough. It is only if one understands what the problem is that one can make further progress, because only then is it worthwhile to track the flow of information more precisely than 'random variation alternating with natural selection'. Following through the logic of Darwinian information flow, one establishes that it is specifically genes—not individuals, nor species, nor the biosphere—that are varied and selected in the evolutionary process, and that they are selected for the ability to promote their own spread through the population, not for 'strength', 'fitness', or 'benefit' to anything. Hence one comes to understand organisms as vehicles for the replication of genes, not vice versa—and so on. Thus one builds the remarkably fruitful field of science and philosophy known as 'neo-Dawinism'.

In the absence of such an understanding, what Dennett called 'Darwin's dangerous idea' is not 'dangerous'; that is to say, it is not fruitful. Take, for example Steven Jay Gould's idea of evolution through 'spandrels', which he named after the gaps between architectural features such as arches. His point is that no one designs spandrels—their shapes are accidental consequences of the design of those other features—and yet they sometimes come in useful later; and he points out that, analogously, there are many examples in the biosphere of new adaptations evolving from features accidentally created by the evolution of something unrelated. The fact that some feature of an organism can be non-evolved but still useful seems, to Gould, to contradict the neo-Darwinist theory that *all* the usefulness in the biosphere has evolved by adaptation to selection pressures. But this is a misunderstanding. For there are two kinds of usefulness, one of them requiring explanation and the other not. When Paley, in his imagination, found a watch on a heath, he did not wonder at the Providence that created *heaths* useful for losing, and then finding, watches that could be referred to in philosophical arguments. For although the heath was useful for that purpose, it showed no sign of being designed for that use: a differently-shaped heath—or a beach or a street—would do just as well. For that matter, Paley could not, in his celebrated argument, have referred to the night sky instead of either a watch or a living organism, even though the night sky was, at that time, far more useful for accurate time-keeping than any watch. Again, the sky, despite its usefulness, shows no sign of being *adapted* (or designed) to be used: it shows (in Paley's words) no 'arrangement, disposition of parts, subservi-ency of means to an end, relation of instruments to a use'. On the contrary, if the solar system's 'different parts had been . . . placed after any other manner', then in most cases the sky would still keep time just as accurately. The same goes for every other use to which the sky has been put. And the same goes for spandrels, both architectural and biological. Thus, no study of where the know-ledge in biological spandrels comes from, and how it gets into genes, can ever lead anywhere, for there is no knowledge in spandrels.

Furthermore, if one does not understand the role of knowledge creation in evolution, one can easily be led, as Gould was, to deny that evolution has created any objective progress in the biosphere. But it certainly has, for the genes that code for (say) the structure of the eye embody objective knowledge of the laws of optics, while those that code for the human brain embody knowledge of the laws of epistemology. None of that knowledge was present in our unicellular ancestors.

Dawkins' arguments for 'neo-Darwinism' stand by themselves as valid: the theory is true. But truth—correspondence with real-ity—is not a sufficient condition for a good explanation: spandrels are real too. So are atoms, and the theory that 'adaptations are entirely caused by the interactions between atoms' is perfectly true: yet it explains nothing. In fundamental fields, one of the most important features of a good explanation is fruitfulness. My own view is that the connections I have sketched here between evolution and the growth of human knowledge are part of a wider unity that also involves quantum physics and the theory of computation, as described in my book *The Fabric of Reality*. So, for instance, in the multiverse, the precise form of a knowledge-laden structure (such as a gene) is the same across a wide swathe of parallel universes, because any small differences between its structure in different universes tend to be eliminated by the error-correcting processes of natural selection. Other forms of complex information—say, the precise distribution of grains of sand on a beach, or the precise positions of stars in a galaxy—are not subject to such error correction and are therefore different in different universes. Thus the knowledge-laden structures are *big*, in the multiverse, while many of the objects, such as galaxies, that have large-scale structure in any one universe, have little or none in the multiverse.

It is only the 'neo-Darwinist' version of evolution theory that has turned out to illuminate other fields in this way. That is smoking-gun evidence of a good explanation.

DEEP COMMONALITIES
BETWEEN LIFE AND MIND

Steven Pinker

> US television talk-show host Jay Leno, interviewing a
> passerby: How do you think Mount Rushmore was
> formed?
> Passerby: Erosion?
> Leno: Well, how do you think the rain knew to not only
>   pick four presidents—but four of our *greatest* presi-
>   dents? How did the rain know to put the beard on
>   Lincoln and not on Jefferson?
> Passerby: Oh, just luck, I guess.

I AM a cognitive scientist, someone who studies the nature
of intelligence and the workings of the mind. Yet one of
my most profound scientific influences has been Richard
Dawkins, an evolutionary biologist. The influence runs deeper
than the fact that the mind is a product of the brain and the brain
a product of evolution; such an influence could apply to someone
who studies any organ of any organism. The significance of
Dawkins' ideas, for me and many others, runs to his characterization of the very nature of life and to a theme that runs throughout
his writings: the possibility of deep commonalities between life
and mind.

Scientists, unlike literary scholars, are ordinarily not a fitting
subject of exegesis and thematic analysis. A scientist's writings
should be transparent, revealing facts and explanations directly.
Yet I find that Dawkins' ideas repay close reflection and re-
examination, not because he is a guru issuing enigmatic pro-
nouncements for others to ponder, but because he continually

engages the deepest problems in biology, problems that continue to challenge our understanding.

When I first read Dawkins I was immediately gripped by concerns in his writings on life that were richer versions of ones that guided my thinking on the mind. The parallels concerned both the content and the practice of the relevant sciences.

The first critical theme is an attention to adaptive complexity as the paramount phenomenon in need of an explanation, most forcibly expressed in *The Blind Watchmaker* and *Climbing Mount Improbable*. In the case of life, we have the remarkable adaptations of living things: echolocation, camouflage, the vertebrate eye, and countless other 'organs of extreme perfection and complication', in Darwin's words, which represent solutions to formidable engineering problems. In the case of mind, we have the remarkable powers of human cognition: the ability to recognize objects and materials, plan and execute movement, reason and remember, speak and understand.

I shared, moreover, Dawkins' impatience with fellow scientists who provide passable accounts of relatively peripheral aspects of their subject matter, but who, when it came to mechanistic explanations for adaptive complexity, were too easily satisfied with verbal formulae and vague hand-waving. Dawkins did not disguise his exasperation with Stephen Jay Gould's claims that he had revolutionized the theory of evolution with addenda such as punctuated equilibrium, species-selection, and exaptation. But these addenda, Dawkins pointed out, did not address the main problem of adaptive complexity in life and so left the core of the theory of natural selection (which does solve the problem) untouched. Many cognitive scientists, I often grumble, also seem to content themselves with verbal substitutes for explanatory mechanisms, such as 'strategies', 'general intelligence', 'plasticity', or 'extracting regularities'.

The discomfort with inadequate explanations of key phenomena underlies another area of resonance—the conviction that in some areas of science there is an indispensable role for the exploration of ideas, their logical adequacy, and their explanatory power, rather than equating science with the obsessive gathering of data. Biology today, especially molecular biology, is massively weighted toward laboratory work, and any hint of theory is considered scholastic or archaic. In the case of molecular biology this attitude is particularly amnesic, because at the dawn of the field in the 1940s there was an obsession with the theoretical preconditions constraining any putative candidate for the machinery of life (as expressed, for example, in the influential treatise 'What is Life?' by Erwin Schrödinger, a theoretical physicist).

Dawkins has been unapologetic in insisting that a complete biology must lay out the implications of its theories, perhaps most forcibly in his essay 'Universal Darwinism', which audaciously argued that natural selection is not only the best theory of the evolution of life on earth, but almost certainly the best theory of the evolution of life anywhere in the universe. I believe that in cognitive science, too, the demands on adequate theories are so stringent as to carve out an essential place for theoretical analysis. In Dawkins' case, this encourages a blurring of his writing for his fellow scientists and for informed nonspecialists: his more popular books certainly cannot be considered 'popularization', nor is his most technical book, *The Extended Phenotype*, restricted to specialists. This is an example I try to emulate.

A second major theme in Dawkins' writings on life that has important parallels in the understanding of the mind is a focus on *information*. In *The Blind Watchmaker* Dawkins wrote, 'If you want to understand life, don't think about vibrant, throbbing gels and oozes, think about information technology.' Dawkins has tirelessly emphasized the centrality of information in biology—the storage of genetic information in DNA, the computations embodied in transcription and translation, and the cybernetic feedback loop that constitutes the central mechanism of natural selection itself, in which seemingly goal-oriented behavior results from the directed adjustment of some process by its recent consequences. The centrality of information was captured in the metaphor in Dawkins' book title *River Out of Eden*, the river being a flow of information in the generation-to-generation copying of genetic material since the origin of complex life. It figured into his *Blind Watchmaker* simulations of the evolutionary process, an early example of the burgeoning field of artificial life. It also lies behind his influential theory of memes, which illustrates that the logic of natural selection applies to any replicator which carries information with a certain degree of fidelity. Dawkins' emphasis on the ethereal commodity called 'information' in an age of biology dominated by the concrete molecular mechanisms is another courageous stance. There is no contradiction, of course, between a system being understood in terms of its information content and it being understood in terms of its material substrate. But when it comes down to the deepest understanding of what life is, how it works, and what forms it is likely to take elsewhere in the universe, Dawkins implies that it is abstract conceptions of information, computation, and feedback, and not nucleic acids, sugars, lipids, and proteins, that will lie at the root of the explanation.

All this has clear parallels in the understanding of the mind. The 'cognitive revolution' of the 1950s, which connected psychology with the nascent fields of information theory, computer science, generative linguistics, and artificial intelligence, had as its central premise the idea that knowledge is a form of information, thinking a form of computation, and organized behavior a product of feedback and other control processes. This gave birth to a new science of cognition that continues to dominate psychology today, embracing computer simulations of cognition as a fundamental theoretical tool, and the framing of hypotheses about computational architecture (serial versus parallel processing, analogue versus digital computation, graphical versus list-like representations, etc.) as a fundamental source of experimental predictions. As with biology, an emphasis on information allows one to discuss cognition in a broader framework from the particular species found on earth, extending to the nature of processes we would wish to consider intelligent anywhere in the universe. And, as in biology, an emphasis on information unfortunately must withstand a strong current toward experimental studies of physical mechanisms (in this case the physiology of the brain) accompanied by a mistrust of theory and analysis. Again there is no contradiction between studying information processing systems and studying their physical implementation, but there has been a recent tendency to downplay the former, at a cost of explanatory adequacy.

The parallel use of information-theoretic concepts in biology and cognitive science (particularly linguistics) is no secret, of course, and is evident in the reliance of genetics on a vocabulary borrowed from linguistics. DNA sequences are said to contain letters and punctuation, may be palindromic, meaningless, or synonymous, are transcribed and translated, and are even stored in libraries. Biologists occasionally describe development and physiology as following rules, most notably in the immunologist Niels Jerne's concept of the 'generative grammar of the immune system'.

A final shared theme in life and mind made prominent in Dawkins' writings is the use of mentalistic concepts in biology, most boldly in his title *The Selfish Gene*. The expression evoked a certain amount of abuse, most notoriously in the philosopher Mary Midgley's pronouncement that 'Genes cannot be selfish or unselfish, any more than atoms can be jealous, elephants abstract or biscuits teleological' (a throwback to the era in which philosophers thought that their contribution to science was to educate scientists on elementary errors of logic encouraged by their sloppy use of language). Dawkins' main point was that one can understand the logic of natural selection by imagining that the genes are agents executing strategies to make more copies of themselves. This is very different from imaging natural selection as a process that works toward the survival of the group or species or the harmony of the ecosystem or planet. Indeed, as Dawkins argued in *The Extended Phenotype*, the selfish-gene stance in many ways offers a more perspicuous and less distorting lens with which to view natural selection than the logically equivalent alternative in which natural selection is seen as maximizing the inclusive fitness of individuals. Dawkins' use of intentional, mentalistic expression was extended in later writings in which he alluded to animals' knowing or remembering the past environments of their lineage, as when a camouflaged animal could be said to display a knowledge of its ancestors' environments on its skin.

The proper domain of mentalistic language, one might think, is the human mind, but its application there has not been without controversy either. During the reign of behaviorism in psychology in the middle decades of the twentieth century, it was considered as erroneous to attribute beliefs, desires, and emotions to humans as it would be to genes, atoms, elephants, or biscuits. Mentalistic concepts, being unobservable and subjective, were considered as unscientific as ghosts and fairies and were to be eschewed in favor of explaining behavior directly in terms of an organism's current stimulus situation and its past history of associations among stimuli and rewards. Since the cognitive revolution, this taboo has been lifted, and psychology profitably explains intelligent behavior in terms of beliefs and desires. This allows it to tap into the world of folk psychology (which still has more predictive power when it comes to day-to-day behavior than any body of scientific psychology) while still grounding it in the mechanistic explanation of computational theory.

In defending his use of mentalistic language in biological explanation, Dawkins has been meticulous in explaining that he does not impute conscious intent to genes, nor does he attribute to them the kind of foresight and flexible cleverness we are accustomed to in humans. His definitions of 'selfishness', 'altruism', 'spite', and other traits ordinarily used for humans is entirely behavioristic, he notes, and no harm will come if one remembers that these terms are mnemonics for technical concepts and heuristics for generating predictions rather than direct attributions of the human traits.

I sometimes wonder, though, whether caveats about the use of mentalistic vocabulary in biology are stronger than they need to be—whether there is an abstract sense in which we can *literally* say that genes are selfish, that they try to replicate, that they know about their past environments, and so on. Now of course we have no reason to believe that genes have conscious experience, but a dirty secret of modern science is that we have no way of explaining the fact that *humans* have conscious experience either (conscious experience in the sense of raw first-person subjective awareness—the distinction between conscious and unconscious processes, and the nature of self-consciousness, are entirely tractable scientific topics). No one has really explained why it *feels like something* to be a hunk of neural tissue processing information in certain complex patterns. So even in the case of humans, our use of mentalistic terms does not depend on a commitment on how to explain the subjective aspects of the relevant states, but only on their functional role within a chain of computations.

Taking this to its logical conclusion, it seems to me that if information-processing gives us a good explanation for the states of knowing and wanting that are embodied in the hunk of matter called a human brain, there is no principled reason to avoid attributing states of knowing and wanting to other hunks of matter. To be specific, nothing prevents us from seeking a generic characterization of 'knowing' (in terms of the storage of usable information) that would embrace both the way in which people know things (in their case, in the patterns of synaptic connectivity in brain tissue) and the ways in which the genes know things (presumably in the sequence of bases in their DNA). Similarly, we could frame an abstract characterization of 'trying' in terms of negative feedback loops, that is, a causal nexus consisting of repeated or continuous operations, a mechanism that is sensitive to the effects of those operations on some state of the environment, and an adjustment process that alters the operation on the next iteration in a particular direction, thereby increasing the chance that that aspect of the environment will be caused to be in a given state. In the case of the human mind, the actions would be muscle movements, the effects would be detected by the senses, and the adjustments would be made by neural circuitry programming the next iteration of the movement. In the case of the evolution of genes, the actions would be extended phenotypes, the effects would be sensed as differential mortality and fecundity, and the adjustment would be made in terms of the number of descendants resulting in the next generation.

This characterization of beliefs and desires in terms of information rather than physical incarnation may overarch not only life and mind but other intelligent systems such as machines and societies. By the same token it would embrace the various forms of intelligence implicit in the bodies of animals and plants, which we would not want to attribute either to fully human cogitation nor to the monomaniacal agenda of replication characterizing the genes. When the coloration of a viceroy butterfly fools the butterfly's predators by mimicking that of a more noxious monarch butterfly, there is a kind of intelligence being manifest. But its immediate goal is to fool the predator rather than replicate the genes, and its proximate mechanism is the overall developmental plan of the organism rather than the transcription of a single gene.

In other words the attribution of mentalistic states such as knowing and trying can be hierarchical. The genes, in order to effect their goal of making copies of themselves, can help build an organ whose goal is to fool a predator. The human mind is another intelligent mechanism built as part of the intelligent agenda of the genes, and it is the seat of a third (and the most familiar) level of intelligence: the internal simulation of possible behaviors and their anticipated consequences that makes our intelligence more flexible and powerful than the limited forms implicit in the genes or in the bodies of plants and animals. Inside the mind, too, we find a hierarchy of subgoals (to make a cup of coffee, put coffee grounds in the coffeemaker; to get coffee grounds, grind the beans; to get the beans, find the package; if there is no package, go to the store; and so on). Computer scientists often visualize hierarchies of goals as a stack, in which a program designed to achieve some goal often has to accomplish a subgoal as a means to its end, whereupon it 'pushes down' to an appropriate subroutine, and then 'pops' back up when the subroutine has accomplished the subgoal. The subroutine, in turn, can call a subroutine of its own to accomplish an even smaller and more specialized subgoal. (The stack image comes from a memory structure that keeps track of which subroutine called which other subroutine, and works like a spring-loaded stack of cafeteria trays.) In this image, the best laid plans of mice and men are the bottom layers of the stack, and above them is the intelligence implicit in their bodies and genes, with the topmost goal being the replication of genes that makes up the core of natural selection.

It would take a good philosopher to forge bulletproof characterizations of 'intelligence', 'goal', 'want', 'try', 'know', 'selfish', 'think', and so on, that would embrace minds, robots, living bodies, genes, and other intelligent systems. (It would take an even better one to figure out how to reintroduce subjective experience into this picture when it comes to human and animal minds.) But the promise that such a characterization is possible—that we can sensibly apply mentalistic terms to biology without shudder quotes—is one of Dawkins' legacies. If so, we would have a substantive and deep explanation of our own minds, in which parochial activities like our own thinking and wanting would be seen as manifestations of more general and abstract phenomena.

The idea that life and mind are in some ways manifestations of a common set of principles can enrich the understanding of both. But it also mandates not confusing the two manifestations—not forgetting what it is (a gene? an entire organism? the mind of a person?) that knows something, or tries something, or wants something, or acts selfishly. I suspect that the biggest impediment to accepting the insights of evolutionary biology in understanding the human mind is in people's tendency to confuse the various entities to which a given mentalistic explanation may be applied.

One example is the common tendency to assume that Dawkins' portrayal of 'selfish genes' implies that organisms in general, and people in particular, are ruthlessly egoistic and self-serving. In fact nothing in the selfish-gene view predicts that this should be so. Selfish genes are perfectly compatible with selfless organisms, since the genes' goal of selfishly replicating themselves can be implemented via the sub-goal of building organisms that are wired to do unselfish things such as being nice to relatives, extending favors in certain circumstances, flaunting their generosity under other circumstances, and so on. (Indeed much of *The Selfish Gene* consists of explanations of how the altruism of organisms is a consequence of the selfishness of genes.) Another example of this kind of confusion is the common claim that sociobiology is refuted by the many things people do that don't help to spread their genes, such as adopting children or using contraception. In this case the confusion is between the motive of genes to replicate themselves (which does exist) and the motive of people to spread their genes (which doesn't). Genes effect their goal of replication via the sub-goal of wiring people with certain goals of their own, but replication per se need not be among those sub-sub-goals: it's sufficient for people to seek sex and to nurture their children. In the environment in which our ancestors were selected, people pursuing those goals automatically helped the relevant genes pursue theirs (since sex tended to lead to babies), but when the environment changed (such as when we invented contraception) the causal chain that used to make sub-goals bring about superordinate goals no longer were in operation.

I suspect that these common fallacies arise from applying a Freudian mindset to evolutionary psychology. People conceive of the genes as the deepest, truest essence of a person, the part that harbors his deepest wishes, and think of conscious experience and overt behavior as a superficial veneer hiding these ulterior motives. This is a fallacy because the motives of the genes are entirely different from the motives of the person—they are a play within a play, not the interior monologue of a single cast of players.

More generally, I think it was the ease of confusing one level of intelligence with another that led to the proscription of mentalistic terms in behaviorism and to the phobia of anthropomorphizing organisms or genes in biology. But as long as we are meticulous about keeping genes, organisms, and brains straight, there is no reason to avoid applying common explanatory mechanisms (such as goals and knowledge) if they promise insight and explanation.

The promise of applying common tools to life and mind, and the danger in failing to distinguish which is the target of any particular explanation, can also, I think, be seen in discussions of the relevance of memes to human mind and culture. Dawkins has suggested that his discussion of memes was largely meant to illustrate the information-theoretic nature of the mechanism of natural selection—that it was not particular to DNA or carbon-based organisms or life on earth but applied to replicators of any sort. Others have treated his suggestions about memes as an actual theory of cultural change, some cleaving to a tight analogy between the selection of genes and the selection of memes, others exploring a larger family of models of cultural evolution, epidemiology, demographics, and gene-culture coevolution, I think that the mind-life parallel inherent in memetics holds out the promise of new ways of understanding cultural and historical change, but that it also poses a danger.

Many theorists, partly on the basis of Dawkins' arguments about the indispensability of natural selection in explaining complex design in living things, write as if natural selection, applied to memes rather than genes, is the only adequate explanation of complex design in human cultural achievements. To bring culture into biology, they reason, one shows how it evolved by its own version of natural selection. But that doesn't follow, because the products of evolution don't have to *look like* the process of evolution. In the case of cultural evolution they certainly don't look alike—human cultural products are not the result of an accumulation of copying errors, but are crafted through bouts of concerted brainwork by intelligent designers. And there is nothing in Dawkins' Universal Darwinism argument that makes this observation suspect. While it remains true that the origin of complex design on earth requires invoking selection (given the absence of any alternative mechanisms adequate to the task), in the case of complex design in culture we do have an alternative, namely the creative powers of the human brain. Ultimately we have to explain the complexity of the brain itself in terms of genetic selection, but then the ladder can be kicked away and the actual process of cultural creation and transmission studied without prejudice.

A final connection. Religion has become a major theme of Dawkins' recent writings, and here too life and mind figure into the argument in related ways. The appearance of complex design in the living world was, of course, a major argument for belief in God throughout history, and a defensible one before it was undermined by the ability of natural selection to generate the appearance of design without a designer. As Dawkins wrote in *The Blind Watchmaker*, 'Although atheism might have been logically tenable before Charles Darwin, Darwin made it possible to be an intellectually fulfilled atheist.' I believe that a parallel development has taken place with regard to beliefs about the mind. The complexity of human intelligence strikes many people as compelling evidence for the existence of a soul in the same way that the complexity of life was seen as evidence for the existence of a designer. Now that intelligence may be explicable in material terms, as a kind of information processing in neural circuitry (with the circuitry itself explicable by natural selection), this second source of intuitive support for spiritual beings is being undermined. Just as evolutionary biology made it possible for intellectually fulfilled people to do without creationism, computational cognitive science makes it possible for them to do without dualism.

ANTIPHONAL VOICES

RICHARD DAWKINS AND THE PROBLEM OF PROGRESS

Michael Ruse

> Directionalist common sense surely wins on the very long
> time scale: once there was only blue-green slime and now
> there are sharp-eyed metazoa.[1]

ONE of the many attractive things about the writings of
Richard Dawkins is his willingness to state his positions
clearly and forcefully. No hiding of ideas in ambiguity or
of saying one thing in the text and then qualifying it to death with
a thousand footnotes. In the language of the Bible, Dawkins lets
his yea be yea and his nay be nay. Nowhere has Dawkins been
more forthright than in his endorsement of the idea of evolutionary progress. He believes in it, he has said so many times, and he
has argued for it.

Yet, perhaps surprisingly, this is a controversial position, with
today's evolutionists split down the middle on the issue. The
entomologist and sociobiologist Edward O. Wilson is as enthusiastic about progress as is Dawkins. 'Progress, then, is a property
of the evolution of life as a whole by almost any conceivable
intuitive standard, including the acquisition of goals and intentions in the behavior of animals.'[2] Stephen Jay Gould, the only
man (apart from Darwin himself) to have competed with Dawkins'
supreme brilliance as a popular writer about evolution, was
adamantly opposed to progress, speaking of it as 'a noxious, cul-
turally embedded, untestable, nonoperational, intractable idea
that must be replaced if we wish to understand the patterns of
history'.[3] It is a delusion engendered by our refusal to accept our
insignificance when faced with the immensity of time.[4]

Let us ask some questions. First, what, if anything, has been the relationship between evolutionary thinking and thoughts of progress? Are we faced with an old dispute that we evolutionists, who take seriously the belief that the past is the key to the present, should consider? Second, why would the notion of progress be controversial? Third, what is Dawkins' position on the idea of progress and is it new? Fourth, is Richard Dawkins right?

The idea of progress is a child of the eighteenth century, the Age of the Enlightenment.[5] In the cultural realm, progress was the belief or conviction that things (education, living standards, knowledge) are getting better, and that we humans are the force behind the improvement.[6] It is often thought to be an idea opposed to Christianity, but it is better to say that it is an idea opposed to the Christian belief in Providence, the idea that only through God's grace can we expect real advance. In the biological realm, and everybody back then was quite explicit that analogy was being drawn with culture, progress meant that among organisms there is an order from simple to complex, from the least to the most, from (as was often said) the monad to the man. (Some put plants at the bottom, some put plants on a different scale.)

Organic evolution came into being on the back of biological progress. The early evolutionists, Denis Diderot[7] and then Jean Baptiste de Lamarck[8] in France, Erasmus Darwin (1794–1796) in England, Johann Wolfgang von Goethe in Germany,[9] were all ardent progressionists in culture and biology, and saw their evolutionism as part and parcel of this general picture.

The story of evolution and progress continued through the nineteenth century from beginning to end. The notorious pre-(Charles) Darwinian work, *The Vestiges of the Natural History of Creation*,[10] published anonymously but later revealed to be the work of the Scottish publisher and author Robert Chambers, was explicit in its progressionism. Later in the century, the philosopher, sociologist, and general man of science Herbert Spencer was the progressionists' progressionist.

The twentieth century likewise saw much enthusiasm for progress, if not always in culture then certainly in biology. Many would argue that, after the *Origin of Species*,[11] the most important book in the history of evolutionary thought is Ronald Fisher's *The Genetical Theory of Natural Selection*.[12] It is a hymn to evolutionary progress, except towards the end where it gives a dire warning that such progress is failing in humans and needs our attention. The central idea, the Fundamental Theorem of Natural Selection, is, by Fisher's explicit admission, intended to push organisms up to an ever-higher peak, thus combating the destructive forces of the Second Law of Thermodynamics.

Although few were as hard-line as Fisher on all of this, with only occasional exceptions we find that the great evolutionists of the age—Julian Huxley, J. B. S. Haldane, Sewall Wright, Theodosius Dobzhansky, George Gaylord Simpson, the botanist G. Ledyard Stebbins, and the ornithologist and systematist Ernst Mayr—agreed that evolution has such an upwards drive.[13]

Why then is the notion of biological progress so controversial? There are two main reasons, one general and one more specific. At the general level, it has been the trend of science—certainly since the Scientific Revolution—to eliminate values.[14] Science tells you about the world of experience, and, in the words of one of America's most famous sports commentators (Howard Cosell), the best science is there to 'tell it like it is'. Karl Popper put it nicely.[15] Science is 'knowledge without a knower'. By this, Popper did not mean that science was not known by people. What he meant was that good science transcends the individual. It is objective. Hence a notion like Jewish science or feminist science is not so much wrong as incoherent.

If science is trying to map reality, then since reality is neither good nor bad—it just is—science itself can have no real values. Or, let us qualify this a bit. Science can have no absolute values, like 'The English are the best nation on earth'. It can have relative values like 'The English today live longer than the English of the Victorian era'. We talk about one sample of water being hotter than another, meaning that on a scale of one to a hundred, one sample scores higher than the other. But we are not saying that it is better overall to be hotter. We ourselves might think that in certain circumstances it is better to be hotter—we might think, for instance when we are making tea, that in every circumstance it is better—but that is a value we impute to the world rather than one we find there.

The trouble is that progress does seem to be a value notion and not obviously a relative-value notion. Many people would agree that humans seem in some absolute sense better than warthogs, and they in turn seem absolutely better than the AIDS virus (or the equivalents in the distant past). In the light of this general point, we can now go on to ask whether there are specific reasons why biological progress is problematical. And the answer is that there are such reasons in the history of evolutionary theorizing to think that, while it was surely the case that at the beginning evolution and progress were joined like Siamese twins, two major acts of surgery have separated them, or rather *should* have separated them.

First there was Charles Darwin's mechanism of natural selection. It is not a tautology as critics often claim—the fittest survive and by definition the fittest are the survivors—but it is relativistic. What might make for survival in one case does not necessarily make for survival in another. Suppose, for some population, being white skinned has adaptive virtues over being black skinned. Suppose that the major selective force is predation, or some such thing, and white protects more than black. Change predators, and it could well be that being white might be a handicap and being black might be a virtue. Given this relativism, it is hard to see why one feature or set of features might be promoted over all others, in some absolute way, as the best.

Second there was Mendelian genetics. It is a key part of this theory, and its molecular successors, that the building blocks of evolution, the mutations, are random—not in the sense of being uncaused or even unquantifiable, but in the sense of not appearing to order. You may need white, but you are as likely to get black, pink, or yellow polka dots. It is this fact that distinguished Mendelism—and hence modern evolutionary thinking—from all others. There is no built-in direction to evolutionary change. Such direction as there is has to come from selection, and that again throws us back on relativism. Apparently, there can be no absolute progress.

It would seem, then, that biological progress today is a non-starter. Enter Richard Dawkins, the co-author of the splendid line with which I preface this essay. Dawkins is a Darwinian, ardently so, and for a Darwinian—and speaking as a fellow traveller, I endorse this thinking entirely—the key to understanding the biological world is adaptation, beginning, middle, and end. That certainly does not imply that one will be a progressionist—we have just seen reasons to think that one will not be a progressionist (E. B. Ford was an ultra-adaptationist but not a progressionist)—but if one is into progress, then adaptation will be at the heart of one's definition. As it is for Dawkins. Probably his best statement of his progressionist position came in a review that he wrote of an anti-progressionist book by Gould. Dawkins gave the following definition of progress:

A tendency for lineages to improve cumulatively their adaptive fit to their particular way of life, by increasing the numbers of features which combine together in adaptive complexes.[16]

You might think this a little bit wishy-washy, because the characterization seems to say little or nothing about that all-important creature, *Homo sapiens*. But I think Dawkins here is in much the same position as Darwin in the *Origin*. There, it will be remembered, Darwin steered clear of the man issue, because he wanted first to get his main ideas on the table. It was not at all because he thought man an exception. Here, Dawkins likewise wanted to get his main ideas on the table. It was not at all because he thought his definition has nothing to say about our species. In his great popular overview of modern evolutionary thinking, *The Blind Watchmaker*,[17] Dawkins refers to Harry Jerison's notion of an Encephalization Quotient,[18] this being a kind of universal animal IQ, that works from brain size and subtracts the gray matter simply needed to get the body functioning—whales necessarily have bigger brains than shrews, because they have bigger bodies. What counts is what is left when you take off the body-functioning portion. Thus measured, humans come way out on top, leading Dawkins to reflect: 'The fact that humans have an EQ of 7 and hippos an EQ of 0.3 may not literally mean that humans are 23 times as clever as hippos!' But, he concludes, it does tell us 'something'.[19]

Elsewhere,[20] Dawkins has tied in his thinking about progress with the notion of the 'evolution of evolvability'. Sometimes, you just get evolutionary breakthroughs—like the eukaryotic cell— that have more potential, and hence evolution has made a jump to a new dimension.

Notwithstanding Gould's just scepticism over the tendency to label each era by its newest arrivals, there really is a good possibility that major innovations in embryological technique open up new vistas of evolutionary possibility and that these constitute genuinely progressive improvements.[21] The origin of the chromosome, of the bounded cell, of organized meiosis, diploidy and sex, of the eucaryotic cell, of multi-cellularity, of gastrulation, of molluscan torsion, of segmentation— each of these may have constituted a watershed event in the history of life. Not just in the normal Darwinian sense of assisting individuals to survive and reproduce, but watershed in the sense of boosting evolu-tion itself in ways that seem entitled to the label progressive. It may well be that after, say, the invention of multicellularity, or the invention of metamerism, evolution was never the same again. In this sense, there may be a one-way ratchet of progressive innovation in evolution.[22]

Dawkins has always made brilliant use of metaphor—selfish gene, blind watchmaker, mount improbable—and metaphor is much involved in the thinking about progress. In *The Blind Watchmaker*, the metaphor of bigger and bigger on-board computers (aka brains) plays a vital role, as it has elsewhere.

Computer evolution in human technology is enormously rapid and unmistakably progressive. It comes about through at least partly a kind of hardware/software coevolution. Advances in hardware are in step with advances in software. There is also software/software coevolution. Advances in software make possible not only improvements in short-term computational efficiency—although they certainly do that—they also make possible further advances in the evolution of the software. So the first point is just the sheer adaptedness the advances of software make for efficient computing. The second point is the progressive thing. The advances of software open the door—again I wouldn't mind using the word 'floodgates' in some instances—open the floodgates to further advances in software.[23]

Evolution is cumulative, for it has 'the power to build new progress on the shoulders of earlier generations of progress'. And brains, especially the biggest and best brains, are right there at the heart, or (perhaps we should say) end: 'I was trying to suggest by my analogy with software/software coevolution, in brain evolution that these may have been advances that will come under the heading of the evolution of evolvability in [the] evolution of intelligence.'

I have praised Dawkins for being forthright, and for avoiding the scholar's inclination to be bold in the text and then overly cautious in the footnotes. I do not mean to imply that Dawkins' thinking is not nuanced. He is fully aware that the notion of progress is something to be approached with care. Against Gould, although he stated flatly that 'evolution turns out to be clearly and importantly progressive in the short to medium term', he is more guarded when it comes to bigger things, saying only that evolution 'is *probably* progressive in the long term also'.[24] One suspects however that this is a high probability, for against the American's attempt to belittle the notion, Dawkins responds with vigour that 'Gould's attempt to reduce all progress to a trivial, baseball-style artifact constitutes a surprising impoverishment, an uncharacteristic slight, an unwonted demeaning of the richness of evolutionary processes'.[25]

Dawkins, however, is sensitive to the values issue. Often, those who want to get around values—one organism is better than another—try to find some feature that increases up through the history of life. This we might value, especially if we humans just so happen to have the feature more than others, but in itself the feature has no ultimate value. Brain size and genome size have been candidates in the past, neither of which has stood the erosion of time—at least, neither of which makes humans winners! Another popular notion is complexity, and in a sense Dawkins is sympathetic to some version of this. Starting with ideas in information theory, he thinks that more complex organisms would require physically longer descriptions than less complex organisms.

We have an intuitive sense that a lobster, say, is more complex (more 'advanced', some might even say more 'highly evolved') than another animal, perhaps a millipede. Can we *measure* something in order to confirm or deny our intuition? Without literally turning it into bits, we can make an approximate estimate of the information contents of the two bodies as follows. Imagine writing the book describing the lobster. Now write another book describing the millipede down to the same level of detail. Divide the word-count in the one book by the word-count in the other, and you will have an approximate estimate of the relative information content of lobster and millipede. It is important to specify that both books describe their respective animals 'down to the same level of detail'. Obviously, if we describe the millipede down to cellular detail, but stick to gross anatomical features in the case of the lobster, the millipede would come out ahead.

But if we do the test fairly, I'll bet the lobster book would come out longer than the millipede book.[26]

Generally, however, Dawkins is scornful of attempts to pin down quantities of progress, in any absolute sense. He rips into such notions, concluding: 'I recommend that evolutionary writers should no longer, under any circumstances, use the adjectives "higher" and "lower".'[27] I might add that in saying this, Dawkins is walking in the footsteps of Charles Darwin, who wrote on the fly of his own copy of *Vestiges*, 'never use the terms higher and lower'.

How does Dawkins make his case? How does someone who clearly thinks that science should be value neutral, who accepts natural selection and modern genetics, argue for progress? How does a Darwinian argue for progress? Let us approach this question through the method of differences, comparing Dawkins to others.

Gould was against biological progress. Interestingly, early in his career, he was quite sympathetic to the idea. His early major work, *Ontogeny and Phylogeny* argued that humans had won the evolutionary climb.[28] Then he swung against the idea, when he decided that Darwinism applied to human behaviour (human sociobiology) was morally dangerous. To explain the 'illusion' of progress, Gould therefore argued that it is all an artifact of the nature of things—if you start simple, you have no way to go but up.[29] A drunk will fall off the sidewalk if it is bounded by a wall on the other side. It is not that the drunk intends to fall, but that he will. Similarly, it is not that humans are in any sense biologically superior to trilobites. It is basically a question of their coming later.

Dawkins does not reject this argument. It is just that he thinks it trivial and uninteresting. Belittling of life, too. Animals and plants are just more interesting, more complex, more functioning, more adaptive than drunks falling into the gutter. It is not just that we are complex, but that we are adaptively complex, to use a phrase that Dawkins borrows from John Maynard Smith. So there has to be something more. Edward O. Wilson might seem promising here.

The overall average across the history of life has moved from the simple and few to the more complex and numerous. During the past billion years, animals as a whole evolved upward in body size, feeding and defensive techniques, brain and behavioral complexity, social organization, and precision of environmental control—in each case farther from the nonliving state than their simpler antecedents did.[30]

The trouble with Wilson is that he really does not offer too much by way of justification for his progressionism. He thinks it just occurs. Indeed, in his great *Sociobiology: The New Synthesis*, having spoken of the paradox of social evolution, namely that humans have reversed the trend from more complex to less complex (mammalian sociality generally is less complex than hymenopteran sociality), although he gives reasons for the reversal (like the growth of the brain) Wilson does not really spell out why any of this happens.[31] Causally, it seems rather haphazard, although psychologically and metaphysically to the contrary there seems a strong whiff here of the very kind of progressionism that Dawkins wants to eschew. For Wilson, progress is something that seems bound to happen and moreover an out-and-out value notion—an absolute-value notion—and that, far from being a handicap, is a virtue. It means that Wilson can get moral directives from evolution. What has evolved is good, and what has evolved higher up the ladder is better. Ultimately the supreme principle is: 'cherish humans'. This is why Wilson is so concerned to preserve the rainforests. Without such forests and their biodiversity, he believes that we humans will perish.

Although he would agree with the sentiment—biological diversity is a good thing—the underlying thinking is anathema to Dawkins. He has gone after the Prince of Wales with courtesy but with controlled ferocity.[32] The Prince, too well known for the soft side to his thinking, has argued against genetically modified foods on the grounds that they go against the wisdom of nature.[33] They did not evolve, so they must be bad. In words reminiscent of Thomas Henry Huxley,[34] who likewise took this kind of argument to task, Dawkins shows with logic and scorn that those who argue for the wisdom of nature also argue for smallpox and the AIDS virus and genetically caused diseases. A *reductio ad absurdum* of their very position. There is no teleological upward drive to evolution.

Recently, the Cambridge paleontologist Simon Conway Morris has tried another tack, making explicit a line of thought I suspect that many have at the backs of their minds.[35] He argues that there are kinds of pre-existing niches—water, land, air, culture—and life, as it were, hops up from one to the next. We humans have hopped farthest and so we have come out on top. Conway Morris is a Darwinian, so he adds that he sees the pressure of natural selection as that which forces life up the scale.

If brains can get big independently and provide a neural machine capable of handling a highly complex environment, then perhaps there are other parallels, other convergences that drive some groups towards complexity. Could the story of sensory perception be one clue that, given time, evolution will inevitably lead not only to the emergence of such properties as intelligence, but also to other complexities, such as, say, agriculture and culture, that we tend to regard as the prerogative of the human? We may be unique, but paradoxically those properties that define our uniqueness can still be inherent in the evolutionary process. In other words, if we humans had not evolved then something more-or-less identical would have emerged sooner or later.[36]

As it happens, Conway Morris is a Christian—a conservative Anglican—but I doubt that this is what Dawkins would find objectionable to the position, although he might well wonder if hope is driving the hop. (I would!) My suspicion is that, apart from the troublesome assumption that water, land, air, and culture make for a simple progression—Why is the land necessarily superior to the water? Are dogs superior to whales?—there is the assumption of pre-existing niches. But do these make sense? Dawkins is the author of *The Extended Phenotype*, a book that argues that organisms are involved in their surroundings, their niches, and that often if not always it is difficult to distinguish the two.[37] You have a beaver and you have its lodge. Where does the beaver end and the lodge begin; where does the animal end and its niche begin? From a Darwinian perspective, these are not easy or straightforward questions. If the lodge is as important to the beaver's well-being as the tail, and if the beaver did as much in creating the lodge as the tail, why then make an ontological separation between the two? Conway Morris' thinking is too simplistic.

Where then do we turn for help on Dawkins on progress? He makes no secret of his thinking. The notion of an *arms race* is absolutely crucial. Organisms compete against each other—more precisely, organisms of one group compete against the organisms of another group—and the interaction brings adaptive changes to both sides. Classically, the prey runs a little faster, and then the predator has to run a little faster—or starve. We have something akin to the human notion of an arms race, and as with the human notion, we get improvement. The armour gets a little thicker; the guns get a little stronger. The hare gets a little faster; the fox gets a little faster.

In a way, this is all relative progress—does one really want to say an efficient gun is absolutely good?—and Dawkins, much to his credit, exploits this idea to the scientific full. Alone and with John Krebs he has offered careful and fruitful analysis of the ways in which arms races can and might be expected to function. For instance, he distinguishes between asymmetrical arms races (with different kinds of competitors, like prey and predator) and symmetrical arms races (with similar competitors, as one might get in sexual selection). However, what about some kind of absolute progress? Does one get this from arms races? My suspicion is that Dawkins rather thinks that in the end one does. In the paper co-authored with Krebs he writes that even 'if modern predators are no better at catching prey than Eocene predators were at catching Eocene prey, it does at first sight seem to be an expectation of the arms race idea that modern predators might massacre Eocene prey. And Eocene predators chasing modern prey might be in the same position as a Spitfire chasing a jet'.[38] Interestingly, and confirming my claim about Gould's early thinking, one of those cited in support is none other than Gould himself!

In *The Blind Watchmaker*, a similar line of argument is followed. Arms races lead to better armour. In the twentieth century we saw that ultimately this led to more efficient electronics, computers especially. Just as those with the biggest and most efficient computers seem to be the winners of literal arms races, so it is reasonable to conclude that those animals with the biggest on-board computers are the winners. And these, as we have seen, are humans.

Is this an argument unique to Dawkins? In respects, as he himself admits fully, it is not. He traces it back to Darwin, although obviously (since it is a term of the 1930s) arms races are not referred to by that name.[39] In fact, even if the notion of an arms race is in Darwin—in the *Origin*, that is—the person who made most of the notion of arms races in biology was Julian Huxley. Right from his first little book, *The Individual in the*

Animal Kingdom, written before the First World War, Huxley was likening biological evolution to the competition between nations in preparation for war.[40]

However, fascinatingly, although there was no greater absolute progressionist than Julian Huxley—he really did think that humans are best, that we have won the evolutionary struggle, and that all morality follows from this—he did not want to link arms races to absolute progress! The reason for this was twofold. On the one hand, he thought that arms races lead to specialization and that specialization is the way of the dead end. A horse is very specialized for running on the steppes, but at the expense of ever evolving its hoofs into something else as useful. Humans, Huxley thought, are the ultimate generalists—not necessarily the best at anything, but better than anyone else at everything. On the other hand, Huxley had no need of arms races. By his own admission, he was ever attracted to vitalism—he was a firm enthusiast for the thinking of Henri Bergson—and although he realized that vital forces really have no place in science, he always thought that there is a kind of necessary momentum in evolution taking us upwards.

And what of Darwin himself? In the early editions of the *Origin*, although the progressionism is there to be seen—we have seen it in the passage quoted in the footnote above—it is put in rather cagey terms. Darwin writes of 'that vague yet ill-defined sentiment'. By the third edition (of 1861), finding that everyone was reading him in a progressionist fashion and praising him for it, Darwin relaxed and in a much more explicit way started adding biological progress—a notion to which he was always firmly committed.

The point of interest is that Darwin's notion of progress, unlike Julian Huxley's, makes specialization the foundation of improvement.

If we take as the standard of high organisation, the amount of differentiation and specialisation of the several organs in each being when adult (and this will include the advancement of the brain for intellectual purposes), natural selection clearly leads towards this standard: for all physiologists admit that the specialisation of organs, inasmuch as in this state they perform their functions better, is an advantage to each being; and hence the accumulation of variations tending towards specialisation is within the scope of natural selection.[41, 42]

Obviously I am not claiming that Dawkins is cribbing Darwin. Dawkins' use of the metaphor of an arms race, especially in the electronic sense, is his own, as is the notion of the evolution of evolvability. As it happens, this second idea is close to the notion of what makes for upward change of Julian Huxley and Haldane. But the point is that whereas they think that arms-race perfection leads to sterility, Dawkins (and probably Darwin) think that it might lead to breakthroughs and great potential for new kinds of development. (In the paper co-authored with Krebs, Dawkins rather seems to agree that arms races will lead to specialization. But with Darwin, and against Huxley and Haldane, Dawkins thinks that this does not preclude breakthroughs. It may even be necessary to produce breakthroughs, at least sometimes.)

So what does one say in conclusion at such rich thinking? Has Dawkins finally made the intellectual breakthrough that gives us a sound notion of biological progress, one that works in a kind of absolutist sense, and yet stays away from absolute values? I do not want to be read as saying that such an endeavor is logically impossible. In theory, I see no reason why one should not have some property that has generally improved over time, under the control of natural selection, and that we humans have it more than others. We would read in the absolute values, but they would reflect 'the way it is'.

In practice, I doubt that Dawkins has done this. I want to see more about how one links arms races and their specialization to the evolution of evolvability and its scope for a new general level of change. I also want to see more clearly spelt out the notion of evolvability, and to be convinced that this is not all wisdom after the fact—or, at the risk of irritating Dawkins who berated Gould for using analogies drawn from American sports, to be convinced that this is not all Monday morning quarterbacking. What gives a feature the potential for evolvability? After the fact we might say that it had it, but is this something or some kind of thing that we can isolate independently of actual success? Are we just renaming what we already know to be true?

These are intended as genuine questions, not as rhetorical refutations. In his recent book, *The Ancestor's Tale*, Dawkins takes the way that avoids a lot of the tricky questions.[43] By tracing life back from humans, he openly acknowledges from the first that he is himself privileging humans. He is structuring the book according to his own wishes, not those of nature. I do hope that this is not a sign that Richard Dawkins is now withdrawing from the arena of (what we might call) progress studies. He has given a great deal. I hope he will give more.

ENDNOTES

1 R. Dawkins and J. R. Krebs, 'Arms races between and within species', *Proceedings of the Royal Society of London B*, 205 (1979): 489–511; 508.

2 E. O. Wilson, *The Diversity of Life* (Cambridge, MA: Harvard University Press, 1992), 187.

3 S. J. Gould, 'On replacing the idea of progress with an operational notion of directionality', in M. H. Nitecki (ed.), *Evolutionary Progress* (Chicago: University of Chicago Press, 1988).

4 S. J. Gould, *Full House: The Spread of Excellence from Plato to Darwin* (New York: Paragon, 1996).

5 M. Ruse, *Monad to Man: The Concept of Progress in Evolutionary Biology* (Cambridge, MA: Harvard University Press, 1996).

6 J. B. Bury, *The Idea of Progress: An Inquiry into its Origin and Growth* (London: Macmillan, [1920] 1924).

7 D. Diderot, *Diderot: Interpreter of Nature* (New York: International Publishers, 1943).

8 J. B. Lamarck, *Philosophie zoologique* (Paris: Dentu, 1809).

9 R. J. Richards, *The Romantic Conception of Life: Science and Philosophy in the Age of Goethe* (Chicago: University of Chicago Press, 2003).

10 R. Chambers, *Vestiges of the Natural History of Creation* (London: Churchill, 1844). See also E. Darwin, *Zoonomia; or, The Laws of*

Organic Life (London: J. Johnson, 3rd edn., [1794–1796] 1801) and E. Darwin, *The Temple of Nature* (London: J. Johnson, 1803).

11 C. Darwin, *The Origin of Species by Charles Darwin: A Variorum Text*, ed. M. Peckham (Philadelphia, PA: University of Pennsylvania Press, 1959).

12 R. A. Fisher, *The Genetical Theory of Natural Selection* (Oxford: Oxford University Press, 1930).

13 See J. S. Huxley, *Evolution: The Modern Synthesis* (London: Allen & Unwin, 1942); J. S. Huxley and J. B. S. Haldane, *Animal Biology* (Oxford: Oxford University Press, 1927); and G. G. Simpson, *The Meaning of Evolution* (New Haven, CT: Yale University Press, 1949).

14 E. Nagel, *The Structure of Science: Problems in the Logic of Scientific Explanation* (New York: Harcourt, Brace and World, 1961).

15 K. R. Popper, *Objective Knowledge* (Oxford: Oxford University Press, 1972).

16 R. Dawkins, 'Human chauvinism: Review of *Full House* by Stephen Jay Gould', *Evolution*, 51/3 (1997): 1015–1020; 1016.

17 R. Dawkins, *The Blind Watchmaker* (New York: Norton, 1986).

18 H. Jerison, *Evolution of the Brain and Intelligence* (New York: Academic Press, 1973).

19 Dawkins, *The Blind Watchmaker* (1986), 189.

20 R. Dawkins, 'The evolution of evolvability', in C. G. Langton (ed.), *Artificial Life* (Redwood City, CA: Addison-Wesley, 1989).

21 Dawkins, 'The evolution of evolvability' (1989) and J. Maynard Smith and E. Szathmáry, *The Major Transitions in Evolution* (New York: Oxford University Press, 1995).

22 Dawkins, 'Human chauvinism: Review of *Full House* by Stephen Jay Gould' (1997), 1019–1020.

23 Ruse, *Monad to Man: The Concept of Progress in Evolutionary Biology* (1996), 469. This is from a presentation given in Melbu, Norway, in 1989.

24 Dawkins, 'Human chauvinism' (1997), 1016 (my italics).

25 Dawkins, 'Human chauvinism' (1997), 1020.

26 R. Dawkins, *A Devil's Chaplain: Reflections on Hope, Lies, Science and Love* (Boston and New York: Houghton Mifflin, 2003), 100.

27 R. Dawkins, 'Progress', in E. F. Keller and E. Lloyd (eds.), *Keywords in Evolutionary Biology* (Cambridge, MA: Harvard University Press, 1992), 263–272.

28 S. J. Gould, *Ontogeny and Phylogeny* (Cambridge, MA: Belknap Press, 1977). See also S. J. Gould, *The Mismeasure of Man* (New York: Norton, 1981).

29 Gould, *Full House: The Spread of Excellence from Plato to Darwin* (1996).

30 Wilson, *The Diversity of Life* (1992), 187.

31 E. O. Wilson, *Sociobiology: The New Synthesis* (Cambridge, MA: Harvard University Press, 1975). See also E. O. Wilson, *The Future of Life* (New York: Vintage Books, 2002).

32 R. Dawkins, 'An Open Letter to Prince Charles', in M. Ruse and D. Castle (eds.), *Genetically Modified Foods* (Buffalo, NY: Prometheus, 2002), 16–19.

33 Charles, Prince of Wales, 'Reith Lecture 2000', in M. Ruse and D. Castle (eds.), *Genetically Modified Foods* (Buffalo, NY: Prometheus, 2002), 11–15.

34 T. H. Huxley, *Evolution and Ethics and other Essays* (London: Macmillan, 1893).

35 S. Conway Morris, *Life's Solution: Inevitable Humans in a Lonely Universe* (Cambridge: Cambridge University Press, 2003).

36 Conway Morris, *Life's Solution: Inevitable Humans in a Lonely Universe* (2003), 196.

37 R. Dawkins, *The Extended Phenotype: The Gene as the Unit of Selection* (Oxford: W. H. Freeman, 1982).

38 Dawkins and Krebs, 'Arms races between and within species' (1979), 490.

39 I have found that Darwin did sometimes use the word 'race' in the sense of competition (as opposed to class) in the first edition of the *Origin*. At least one passage seems to tie things into progress in a very Dawkinsian way, although the word 'race' seems more of a synonym for struggle than quite the idea of a moving competition.

The inhabitants of each successive period of the world's history have beaten their predecessors in the race for life, and are, in so far, higher in the scale of nature; and this may account for that vague yet ill-defined sentiment, felt by many palæontologists, that organisation on the whole has progressed. (Darwin, *The Origin of Species: A variorum Text* (1959), 561)

In the little book on orchids, *On the Various Contrivances by which British and Foreign Orchids are Fertilized by Insects*, published just after the *Origin* in 1862, Darwin got closer to the idea of 'race' as is used in 'arms race'. He wrote:

If the Angræcum in its native forests secretes more nectar than did the vigorous plants sent me by Mr. Bateman, so that the nectary becomes filled, small moths might obtain their share, but they would not benefit the plant. The pollinia would not be withdrawn until some huge moth, with a wonderfully long proboscis, tried to drain the last drop. If such great moths were to become extinct in Madagascar, assuredly the Angræcum would become extinct. On the other hand, as the nectar, at least in the lower part of the nectary, is stored safe from depredation by other insects, the extinction of the Angræcum would probably be a serious loss to these moths. We can thus partially understand how the astonishing length of the nectary may have been acquired by successive modifications. As certain moths of Madagascar became larger through natural selection in relation to their general conditions of life, either in the larval or mature state, or as the proboscis alone was lengthened to obtain honey from the Angræcum and other deep tubular flowers, those individual plants of the Angræcum which had the longest nectaries (and the nectary varies much in length in some Orchids), and which, consequently, compelled the moths to insert their probosces up to the very base, would be fertilised. These plants would yield most seed, and the seedlings would generally inherit longer nectaries; and so it would be in successive generations of the plant and moth. Thus it would appear that there has been a race in gaining length between the nectary of the Angræcum and the proboscis of certain moths; but the Angræcum has triumphed, for it flourishes and abounds in the forests of Madagascar, and still troubles each moth to insert its proboscis as far as possible in order to drain the last drop of nectar. (Darwin, *On the Various Contrivances by which British and Foreign Orchids are Fertilized by Insects* (London: John Murray, 1862), 201–203.

This word is picked up by Darwin's American friend, the botanist Asa Gray. In a letter to Darwin, on 22 March 1863, he wrote: 'Of course we believers in real design make the most of your "frank" and natural terms, "contrivance, purpose", etc., and pooh-pooh your endeavors to resolve such contrivances into necessary results of certain physical processes, and make fun of the race between long noses and long nectarines.' (J. L. Gray, *Letters of Asa Gray* (Boston: Houghton Mifflin, 1844), vol. 2, page 502) Gray himself was not very keen on biological progress, mainly because he was a botanist—he resisted Darwin's efforts to force him to accept progress—but also because he was an evangelical Presbyterian who thought that God's grace was the key to salvation.

40 J. S. Huxley, *The Individual in the Animal Kingdom* (Cambridge: Cambridge University Press, 1912).

41 Darwin, *The Origin of Species: A Variorum Text* (1959).

42 Compare the version of the passage given in endnote 39 with the version that appears in later editions. Specialization becomes explicit and doubts about progress are suppressed. This is from the sixth edition of the *Origin* (1872):

The inhabitants of the world at each successive period in its history have beaten their predecessors in the race for life, and are, in so far, higher in the scale, and their structure has generally become more specialised; and this may account for the common belief held by so many palæontologists, that organisation on the whole has progressed. (Darwin, *The Origin of Species: A variorum Text* 1959), 561)

43 R. Dawkins, *The Ancestor's Tale* (London: Weidenfeld & Nicolson, 2004).

THE NEST'S TALE:
AFFECTIONATE
DISAGREEMENTS WITH
RICHARD DAWKINS

Patrick Bateson

RICHARD and I are sometimes presented as being highly critical of each other. Those who hope for bloody gladiatorial contests are disappointed when they discover that the circles of our interests and beliefs overlap much more extensively than they had believed. Obviously we disagree about some matters, as I shall describe later. Nevertheless, the disagreements are generally teasing and affectionate, since we are old friends, and have not taken on the character of those bitter wrangles that can disfigure the face of academic life.

For my part, I have much to be grateful to Richard for what he has done. Like numerous other academics working in university biology departments I have taught a great many students who were inspired to take our courses at Cambridge because they had read *The Selfish Gene*. Matters that had puzzled Darwin, such as self-sacrifice, were made clear to them and they discovered about conflicts between the sexes and generations where previously they had not appreciated that any existed. The language of genes' intentions employed by Richard helped them to deal with the complicated dynamics of evolution. 'Untestable!', grumbled some hard-nosed colleagues from disciplines concerned with physiological mechanism. But they didn't understand. Such explanations are not meant to be treated in the way usually employed by an experimental scientist; they provide a framework in which we can start to think about phenomena that would otherwise be neglected.

Richard was clearly and deliberately using a device to help understanding when he attributed motives to genes. He obviously did not think that genes have intentions. It is easier for most of us to get our minds round a problem when we can think of a complex system in terms of the way they strive to reach a specific goal. This is not only true in biology. A great nineteenth-century physicist, called William Hamilton (no relation of the sadly deceased biological hero of the same name), formulated a general and widely accepted teleological principle for use in mechanics. It is a powerful way of thinking about systems, the behaviour of which is determined by many factors. To this day forecasters, having to cope with explaining appallingly complex weather systems, make statements like: 'The front is trying to bring rain in from the west'. I was amused to discover that nineteenth-century theologians (but not Hamilton himself) took his principle as proof of the existence of God. I can just imagine the look on Richard's face if a modern day preacher were to say the same about *The Selfish Gene*!

Levels of Selection

Twenty years ago Richard and I were asked to debate with each other at the Institute of Contemporary Arts in London. The audience were disappointed. 'You were too nice to each other', I was told afterwards. As I have already explained, that was because we share a lot of views in common and also genuinely respect each other. Even so, the audience had wanted a more vigorous debate of one of the outstanding issues that lay between us, namely the level of selection in evolution. Richard had initially pinned it at the level of the gene. I wanted to pin it on fully developed characteristics at any level at which differential survival might have occurred. After all, Darwin had used his metaphor of 'natural selection' because he had been impressed by the ways in which plant and animal breeders artificially selected the characters they sought to perpetuate. Nowadays, I am much less happy with the metaphor of selection than I used to be because it implies undue passivity on the part of the organism,[1] but the question of the level of operation remains.

Some years before our debate I had argued that insistence on the gene as the unit of selection was a bit like arguing that even though people buy cars, the units of selection are the great hydraulic presses that stamp out the car bodies and all the other machine tools that make the car's components. The incompleteness of this analogy was, of course, that cars don't make the machine tools for the next generation of cars. Nevertheless, existence of the tools will be perpetuated by the selection of the cars. The tools will only be dismantled when they wear out (which apparently takes a long time) or when the car ceases to sell well; and even then some of them may be retained when components are recombined in a new model. The evolution of the motor car can be expressed in terms of selfish machine tools intent on producing the best possible component regardless of what the other components are like.

Later I tried another analogy. Having told the public for many years that there was no demand for crusty bread, a few supermarkets cautiously offered such bread along with the flabby stuff which was supposed to be so popular. Many people immediately started to buy the crusty bread. The presumed effect of the selection pressure was that the recipe used for making crusty bread proliferated at the expense of the one used for making the alternative. The phrase in the recipe required for making the preferred bread might be regarded as selfish, because it serves to perpetuate itself. That doesn't mean that customers of the supermarket really select the phrase. They select the bread.[2]

In his essay that accompanied mine when the non-debate at the Institute of Contemporary Arts was published, Richard focused on reductionism and genetic determinism (which we agreed about) but he wrote that: 'I should have preferred to spend the time discussing the genuinely interesting, if rather advanced and specialised, disagreement that Patrick Bateson has with me over units of selection . . .'[3] and later: 'There are, of course, many debates and points of disagreement within socio-biology and related disciplines, and some of these are of great interest, for example the argument raised in the accompanying chapter by Patrick Bateson over genes as units of selection.' In fact, he had already responded to my satisfaction a few years earlier when he drew the distinction between 'vehicle selection' and 'replicator survival'.[4] The agents of differential survival and differential reproductive success will usually be characteristics of whole individuals including the structures they make, but they might be characteristics of molecules or symbiotic groups, or the evolvability of taxonomic lineages as Richard argued himself. As things turned out, however, we should have completed the debate twenty years ago because Richard has reverted to his 'gene-selection' position.

Different Languages and the Problems of Translation

When I reviewed *The Selfish Gene* nearly thirty years ago, I raised the concern that Richard's splendid way of thinking about evolution should be used as a reassertion of the crude role of genes in development.[5] I knew perfectly well that when Richard was writing up the work he did for his doctorate at Oxford several years earlier, he expressed as clear an understanding of development process as you could find at the time. If anybody had any doubts in later years, all they had to do was read the second chapter of *The Extended Phenotype*. Yet, many people have continued to think that Richard is a genetic determinist. Why should this be? I believe the answer lies in the language and the similes he sometimes uses. This is an area where a genuine misunderstanding has arisen between us. Recently the journal *Biology and Philosophy* ran three articles by Kevin Laland, J. Scott Turner, and Eva Jablonka discussing the impact on biology of Richard's book, *The Extended Phenotype*, after it was published in 1982. I was much enjoying reading Richard's response until I came to a scarcely veiled attack on my 'obscurantism'. With friends like that who needs enemies! Richard referred to my 'superficially amusing but deeply misleading suggestion that a gene is a nest's way of making another nest'. It related to a passage in my review of *The Selfish Gene*. It is worth quoting the passage since the fidelity of the replication has suffered from Chinese Whispers, about which Richard writes so well, and, more seriously, my intent had been corrupted over time. So much so, indeed, that I too had temporarily forgotten the point that I had wanted to make. What I actually wrote was this:

A legitimate focus on a gene's intentions should not be used as an excuse for resuscitating moribund preformationism. . . . Dawkins accepts all this but then reveals his uncertainty about which language he is using by immediately giving special status back to the gene as the programmer. Consider a case in which the ambient environmental temperature during development is crucial for the expression of a particular phenotype. If the temperature changes by a few degrees the survival machine is beaten by another one. Would not that give as much status to a necessary temperature value as to a necessary gene? The temperature value is also required for the expression of a particular phenotype. It is also stable (within limits) from one generation to the next. It may even be transmitted from one generation to the next if the survival machine makes a nest for its offspring. Indeed, using Dawkins' own style of teleological argument one could claim that the bird is the nest's way of making another nest.

Dawkins' riposte to my tease was that nest material doesn't have the permanence of DNA. Later he developed the point, arguing that nests do not have the causal significance of genes. 'There is a causal arrow going from gene to bird, but none in the reverse direction. A changed gene may perpetuate itself better than its unmutated allele. A changed nest will do no such thing unless, of course, the change is due to a changed gene, in which case it is the gene that is perpetuated, not the nest.'[6]

Richard realized, however, that we might have been at cross purposes and on the next page wrote: 'As is so often the case, an apparent disagreement turns out to be due to mutual misunderstanding. I thought Bateson was denying proper respect to the Immortal Replicator. Bateson thought that I was denying proper respect to the Great Nexus of complex causal factors interacting in development.' His ironic reference to the Great Nexus (not a phrase I have ever used) was not intended to be complimentary. Letting that pass for the moment, a deeper issue is at stake, namely the confusion that arises when a switch is made between different forms of thought.

I was concerned that the power of the selfish gene language was being used to prop up the idea of the gene as 'programmer' but admittedly I distracted attention from this point by my teasing suggestion that the teleological language could be played in different ways. I was not claiming that all the necessary conditions for development could be treated as 'replicators' in biological evolution. Nor was I making the vacuous statement that development is complicated. The central point of that passage in my review of *The Selfish Gene* was to do with the kind of slippage that can occur when language is used loosely. I suspect that Richard believes that he has never been guilty of such slippage for he used the simile of the Necker Cube to refer to the ways it might be possible to move from one type of discourse to another. However, even as clear a thinker as Richard sometimes marches into a linguistic quagmire that causes so much confusion in the minds of others.

Richard is aware that he uses 'gene' in distinctly different ways. For population geneticists, a genetic difference is identified by means of a biochemical, physiological, structural, or behavioural difference between organisms (after other potential sources of difference have been excluded by appropriate procedures). Richard suggested that his move backwards and forwards between the language of gene intentions and the more orthodox language of genetic differences was acceptable because they are simply alternative ways of describing the same thing. To make his point, he described perception of the Necker Cube. The front edges of the line drawing of the cube suddenly flip to the back as we look at them. The lines representing the edges of a cube can be seen as though either the top corner of the cube is facing forwards or it is facing away. Each perceived image of the cube is as real as the other and Richard suggests that, in similar ways, the different pictures of the gene translate backwards and forwards into the other. Both perceptions are equally valid.

At first the Necker Cube analogy seems appealing, but it is not exact because the bodies of thought and evidence on which perceptions are based are different for the two ways in which Richard uses 'gene'. In the technically precise language of population geneticists, a genetic allele must be compared with another from which it differs in its consequences. In selfish-gene language, it stands alone as an entity, absolute in its own right. The perception generated by one meaning of gene does not relate to the same evidence as that generated by the other. Both Richard and I may be selfish, but the difference between us certainly is not. It makes no sense to attribute motives to a comparison.

While my nest talk should not be used in the context of *Darwinian* evolution, it might nevertheless throw up interesting questions about evolution in general that are not otherwise asked. Some aspects of the environment may be stable for a very long time and yet are crucial for the expression of an adaptive phenotype. Many important writers have suggested that changes in those environmental factors can produce dramatic alterations in the characteristics of an organism. This thought lay behind C. H. Waddington's interpretation of his own experiments when, for example, he exposed the larvae of fruit flies to a sudden burst of heat and later some of the adults had abnormal wings. If previous environmental constants, such as the acidity of the sea, start to change, they can be major sources of extinction and, more importantly, they can provide the variation in organisms' characteristics on which subsequent Darwinian evolution can act. This source of evolutionary change must not be confused with the adaptation that follows it, but it would be myopic to dismiss it as unimportant in the broader scheme of things.

That stated, I applaud Richard's clear description of what usually matters in Darwinian evolution. Variants must breed true and with sufficient fidelity so that when one variant survives better or reproduces better than its alternative its characteristics are represented in subsequent generations. It is, as Richard would be the first to admit, a reformulation of Darwin's mechanism for adaptive evolutionary change: variation, differential survival, and onward transmission.

The crucial agents necessary for this evolutionary process of adaptation will generally be genes. Nevertheless, Richard states that we should accept with open arms other agents or processes that might operate in the same way. Recently Matteo Mameli has responded to the challenge and extends the evolutionary mechanism in a disciplined way.[7] Mameli asks us to consider the butterfly that lays its eggs on a particular plant, the leaves of which are then eaten by its offspring caterpillars. The individuals retain through pupation some representation of what they have eaten and when the new generation of adult female butterflies have mated, they lay eggs on the particular species of plant they had eaten before metamorphosis. Occasionally the females lay their eggs on other species of plant. If the leaves should be more nutritious than those of the plant species usually eaten by the insects before metamorphosis, the caterpillars will grow faster and may survive better than their competitors. Consequently, without genetic change the butterfly species switches its plant of choice in the course of Darwinian evolution. In this case the variation lies in the behaviour of the adult female butterflies choosing sites for laying their eggs, the differential survival results from differences in nourishment, and onward transmission to the next generation is achieved by an imprinting-like mechanism.

Richard will probably argue that this and other instances like it are special cases and that they don't seriously detract from the argument that, on the whole, the crucial variation on which the Darwinian evolution depends lies in genetic differences. I would agree with him although I would also agree with Mameli that such cases may seem uncommon because we have not been looking for them. Furthermore, they remind us of the three types of question raised by the Darwinian evolutionary mechanism. What are the developmental and other processes that generate variation in the characteristics of organisms? What are the agents of differential survival and differential reproductive success? What are the necessary conditions for recreating successful characteristics in the next generation?

Simplicity and Complexity

Richard recently had a go at me when he discussed the abuse of the term epigenetics which, he claimed, 'has become associated with obscurantism among biologists'. This is followed by a reference to a footnote which reads: 'I am reminded of a satirical version of Occam's Razor, which my group of Oxford graduates mischievously attributed to a rival establishment: "Never be satisfied with a simple explanation if a more complex one is available." And that reminds me to say that Laland [one of the commentators in the *Biology and Philosophy* issue devoted to *The Extended Phenotype*] has missed the irony in my apparent espousal of Bateson's "Great Nexus of complex causal factors interacting in development".'[8]

The mischievous attribution directed at me actually came from me. About thirty years ago I used to enjoy propounding what I called my three heterodox principles for people working on behaviour. The first principle was 'Treat animals like humans until you have good reason to think otherwise'; the second 'Never use a simple explanation if a more complicated one will do instead'; and the third 'Never use a causal explanation if a teleological one will do instead'. The second one became known self-mockingly as the Cambridge principle because Robert Hinde, the driving intellect at the Sub-Department of Animal Behaviour at Cambridge, used to say repeatedly 'Behaviour *is* complicated'. The third principle became known as 'the Oxford principle' for reasons not unconnected with Richard Dawkins himself. The jokes were half serious, but only half. On the Cambridge principle, everyone will agree that an explanation of an organism's development in terms of it merely getting larger is absurd. The thought that a homunculus inside the sperm head grew and grew until magically it became an airline pilot is not only simple, it is blatantly wrong. The opposite conclusion, that everything is connected to everything else—the Great Nexus—is hopelessly vacuous and is not one I have ever derived or supported. Excluding the middle ground is a tactic that the more radical evolutionary psychologists adopt when they seek to justify their own (simple) views about behavioural and cognitive development. But it is not a rhetorical device that I would expect from Richard—especially when describing the thinking of a friend who, he knows, has spent much of his research life attempting to characterize the underlying rules of behavioural imprinting and the principles of behavioural development. My general concern has been with how the undoubted complexities of development might be made more tractable by uncovering principles that make sense of that complexity.

As I see it, the impulse behind Robert Hinde's easily lampooned phrase 'Behaviour *is* complicated' was a wish to provide an approach that is a precondition for constructing a sensible theory or for deriving a coherent principle. Many powerful voices had urged the behavioural and social sciences to model themselves on the success stories of classical physics or molecular biology. The obvious attractions of producing simple, easily understood explanations has meant unfortunately that crucial distinctions have been fudged in the name of being straightforward and analysis has been focused on single factors in the name of clarity—as has been particularly obvious in studies of behavioural and cognitive development. Little progress is made in the end if the straightforwardness and clarity are illusions. Nobody likes to think that his or her pet principles are constrained. Indeed, a common feature of bolder writers is to make a virtue of this dislike and drive grossly stripped-down explanations all over the place as though these were the attractive and necessary simplifications for which everybody craves.

Being complicated for its own sake has no merit either, but explanations are worthless if they do not bear some relation to real phenomena. Robert Hinde's point was that understanding how the parts relate to each other is a precondition to understanding process and understanding process is the precursor to uncovering principles. Inevitably, tension still exists between those who emphasize differences and focus on complexity and those who unify and simplify. But given that there is no royal road, the old 1960s slogan 'make love, not war' is worth remembering.

Conclusion

Some may argue that essays intended to celebrate friends should not be critical. It is comforting to be praised, and Richard certainly deserves heaps of praise. Even so, constructive criticism should also be seen as flattery and may be more stimulating. This essay raises questions that I hope he will take in the spirit in which they are given. The levels of selection debate that seemed to have been resolved in 1982 needs to be properly engaged once more. The mistakes and misunderstandings that occur in attempts to translate between different types of language need to be recognized. Just because I admire the clarity and brilliance of his writing, I think it is appropriate to identify where Richard might have led others astray by the very gifts that have made him justifiably famous.

Finally, I hope that we can agree that different processes are likely to be involved in biological evolution. Darwinian evolution operates on characters that have developed within a particular set of conditions. If those conditions are stable for many generations then the evolutionary changes that matter will arise in the way that Richard has so clearly and carefully described. Apparent design is produced, even when it is at the end of the long and complicated process of development. But the environment does not cease to be important for evolution just because it remains constant. Change the environment and the outcome of an individual's development may be utterly different. Indeed, if an individual does not inherit its parents' environment along with their genes and other transmittable factors, it may not be well adapted to the conditions in which it now finds itself. But the altered environmental conditions may throw up variation that was previously hidden and from that may spring new lines of evolution. Active choice and active control by the organism together with its own adaptability may all be important additional drivers of evolutionary change. These possibilities do not conflict with the ideas about the evolution of apparent design, which Richard describes so well, but they can explain why sudden changes in direction can and obviously did occur over the long span of biological evolution.

ENDNOTES

1 I have given reasons for considering the active role of animals in changing the direction of evolution in P. Bateson, 'The return of the whole organism', *Journal of Biosciences*, 30 (2005): 31–39.

2 P. Bateson, 'Sociobiology and human politics', in S. Rose and L. Appignanesi (eds.), *Science and Beyond* (Oxford: Blackwell, 1986), 79–99.

3 R. Dawkins, 'Sociobiology: The new storm in a teacup', in S. Rose and L. Appignanesi (eds.), *Science and Beyond* (Oxford: Blackwell, 1986), 61–78.

4 R. Dawkins, 'Replicators and vehicles', in King's College Sociobiology Group (eds.), *Current Problems in Sociobiology* (Cambridge: Cambridge University Press, 1982a), 45–54.

5 P. Bateson [P.G.], 'Book Review: *The Selfish Gene* by Richard Dawkins', *Animal Behaviour*, 26 (1978): 316–318.

6 R. Dawkins, *The Extended Phenotype* (Oxford: W. H. Freeman, 1982).

7 M. Mameli, 'Nongenetic selection and nongenetic inheritance', *British Journal for the Philosophy of Science*, 55 (2004): 35–71.

8 R. Dawkins, 'Extended phenotype—but not too extended. A reply to Laland, Turner and Jablonka', *Biology and Philosophy*, 19 (2004), 377–396.

WHAT'S THE MATTER
WITH MEMES?

Robert Aunger

I'M old enough to be one of the first generation of people brought to the 'gene's-eye-view' of biology by reading *The Selfish Gene*. It changed my outlook on life and has had a profound influence on my subsequent thinking. This transformative experience has been reproduced in many thousands of readers over the past thirty years. Oddly enough, for someone who has studied memes, I don't remember being particularly taken with the final chapter, 'The Long Reach of the Gene', at the time. What transported me then was the profound general outlook the book provided on why social interactions work the way they do. My professional concerns with the meme concept, introduced by Dawkins in that last chapter, came later, when I became interested in understanding cultural change.

A meme, of course, is defined as the fundamental unit of cultural transmission. From an evolutionary perspective, it plays the role in cultural change equivalent to that of the gene in biological change: as the basic unit of inheritance allowing the accumulation of adaptations. The idea is that, like a gene, a meme is a *replicator* (a concept also first defined by Richard Dawkins in *The Selfish Gene*). Genes replicate through the duplication of DNA strands; cultural replication, or the duplication of memes, takes place through the social transmission of information.

Dawkins was not the first scholar to broach the idea that culture might be underpinned by the replication of bits of information. The idea had been in the air for some time, with a variety of linguistic novelties being coined to describe a cultural replicator over the years: 'culturgen', 'mnemotype', 'culturetype', and 'sociogene'. But Dawkins' use of the word 'meme' caught on. Indeed the story of the spread of the meme meme makes a good case history in memetics (the study of memes).

To show the success of his invention, Dawkins did a web search in 1998 in which 'memetic' (used to eliminate possible confusion with the French word 'même') returned 5,000 web pages while 'culturgen' (the main contemporary rival, fashioned by Lumsden and Wilson in their book *Genes, Mind and Culture*)[1] returned only twenty.[2] In the intervening seven years, the comparative advantage of meme has increased dramatically: memetic now appears on over 168,000 web pages, while culturgen (or culture-gen) lags with 537; other alternatives are nearly invisible. The 'meme' has won the contest to be the accepted name for the fundamental unit of culture. (A particularly convincing sign is the fact that Edward O. Wilson uses 'meme' seven times in his book *Consilience*, while mentioning his own coinage, 'culturgen', only once.) The reason could be purely semantic: 'culturgen' is harder to say, while 'meme' easily blossoms into 'memeplex' or 'meme pool'. On the other hand, perhaps the term's success is due to the fact that millions of people have now read *The Selfish Gene*.

The memetic banner has since been carried forward by a growing battalion. Since *The Selfish Gene* was first published, a number of books (by Blackmore, Dennett, Distin, and myself),[3] numerous articles, an electronic journal (the *Journal of Memetics*), and countless web postings and pages have been devoted to developing the meme meme.

However, when he introduced the idea, Dawkins wasn't intending to inspire a new field of speculation and research; he was actually introducing an example of a second replicator, to show that Darwinian replication is not confined to genes alone.[4] He suggested that successful memes, like other replicators, should exhibit three crucial characteristics: fidelity, fecundity, and longevity.[5] Fidelity refers to the ability of a replicator to retain its information content as it passes from mind to mind. Fecundity is a measure of a replicator's power to induce copies of itself to be made. Longevity is less crucial; it only suggests that memes which survive longer have more opportunities to be copied, so the number of their offspring can increase too.

The meme meme seems to have these qualities in spades. Indeed, the meme idea has spread through both 'highbrow' and popular culture, being used in a variety of ways by different disciplines or interest groups. For example, changes in the frequency with which the birds in an area sing elements of their song has been studied by animal behaviourists as a kind of 'population memetics' (by analogy to the study of changes in gene frequencies in population genetics). Similarly, computer scientists have argued that enabling one robot to imitate the behaviour of another ('meme copying') is a way to get robots to develop 'culture'. Inspiring customers to spread good word-of-mouth about their products is also seen by some business writers as an exciting new tactic for increasing sales—a process they believe takes advantage of 'meme power'. Most infamously, perhaps, theologians have reacted to Dawkins' well-known atheistic stance and espousal of religious beliefs as harmful 'mind viruses' with defences of their beliefs—as in the book by John Bowker debating the question *Is God a Virus?*[26]

Nevertheless, no significant body of empirical research has grown up around the meme concept (the birdsong work being the sole, limited exception), nor has memetics made empirically testable propositions or generated much in the way of novel experimental or observational data. In fact the memetic literature remains devoted almost exclusively to theoretical antagonisms, internecine battles, and scholastic elucidations of prior writings on memes. This is typically the sign of a science in search of a subject matter.

Why is memetic science ailing? I think most of the problems have to do with the lack of a useful definition. I would like to spend my time in this brief essay attempting to clarify this basic issue. As we will see, getting specific about the nature of memes leads to questions about whether there is indeed any subject matter for memetics to study.

So just what is a meme? Dawkins famously argued in *The Selfish Gene* that memes could be 'tunes, catch-phrases, clothes fashions, ways of making pots or of building arches'. This definition allowed memes to be found in various kinds of things: inside people's heads, in people's behaviour, and in artefacts. Susan Blackmore agrees with this broad definition in her book *The Meme Machine*, and argues that memes are implicated in the origins of human biology (particularly our large brains), culture (especially language, religion, art) and technology (which has become more and more efficient at copying and multiplying memes in artefacts like books and the World Wide Web).[7] Basically, she sees memes as driving just about every interesting aspect of human evolution. This makes memes very powerful indeed. The problem is that, if memes explain everything, then they explain nothing. This sort of catch-all definition is too broad to be scientifically useful, and, I believe, accounts for memetics being empirically moribund at present.

I have argued, on the contrary, that what makes the meme concept special as an account of cultural evolution is its role as a *replicator* in culture.[8] This is consistent with Dawkins' original objective in positing the existence of memes as a foil to genes.

The replicator concept has been one of Dawkins' lasting contributions to evolutionary theory. However, finding a way to define replication so that it encompasses all of the known replicators—genes, prions, computer viruses, and memes—has been difficult. I have suggested that replication can be defined as a special relationship between a source and a copy such that four conditions hold:[9]

- causation (the source must play some role in bringing about the conditions that lead to a copy being made);
- similarity (the source and copy must resemble each other in relevant respects);
- information transfer (what makes the copy similar to the source must be derived from the source); and
- duplication (the source and copy must coexist for some time).

What does this definition of replication imply about the nature of memes? Does it restrict their definition in a useful way? I believe it does, but it will take a bit of analysis to see why.

Dawkins, like other memeticists, has argued that memes, like other replicators, can exist in many different forms. In effect, replicators are seen as symbolic entities which can morph from one form to another. Dawkins and others tell stories like the following, in which a gene is duplicated in a rather complicated fashion. Imagine a gene sequencing machine has 'decoded' a stretch of DNA into the familiar sequence of Gs, As, Ts, and Cs (e.g. 'GCATACGATA'). This sequence is then printed onto a piece of paper, which is subsequently fed into another machine that reconstructs the same sequence of amino acids that made up the original DNA. That newly-created length of DNA is finally inserted into the nucleus of a cell and begins to function as evolution has designed it to.

In this example, the gene appears to have gone from being a portion of DNA to a sequence of markings on paper, then back to DNA. In effect, one code has been translated into another and back again, with the two different codes being realized in two different physical substrates. There is a one-to-one correspondence between the two coding systems, each of which has only four values, so high fidelity conversion back and forth is not difficult to achieve.

But let's look at this story more closely. There is certainly a causal chain in which information from one stretch of DNA is transferred to another stretch of DNA through the intermediate step of being stored symbolically on paper. Thinking of information in an abstract way suggests that the gene has been converted to a paper form, and that information inheritance has occurred: the crucial information seems to have been passed right down the line from one 'real' gene to another. However, it also appears nonsensical in evolutionary terms to argue that the symbol sequence on paper *is* a gene: the paper form does not conserve the essential features of a gene, its evolved functions.[10] In particular, the sequence of symbols can't produce a protein, or regulate the operation of other genes, no matter what environment the piece of paper is put in. That is because, in a different coding system on a different physical medium, this capability is lost. The symbol string *does*, however, hold information *about* a gene, which is used by specialized machinery as the basis for putting together the proper sequence of amino acids constituting that gene. So DNA → paper → DNA represents a causal chain, but not an evolutionary lineage. This is because a lineage should constitute a sequence of copies, each of which is able to make further copies of things like itself—that is, working replicators, all along the line.[11]

How can we reconcile the fact that there is causation, information transfer, and the duplication of DNA in this sequence but no evolutionary lineage? The only replication condition this example fails, according to our definition of replication above, is similarity: one copy must be 'like' the next. Is this lack of similarity due to the change in code from DNA to paper? Actually, there is a change in code during 'normal' DNA replication and expression: the duplication and transcription of DNA strands involve RNA (as primers or messengers, respectively); but RNA works via a slightly different coding scheme than DNA (changing one of the four nucleotides). There are also cases in which the replication of cultural information involves code-switching; for example, one can change codes of music from MP3 to WMA or other forms while duplicating files. But all of these music codes are digital (two symbols only) and exist on the same medium: magnetic memory in a music player or computer.

So it isn't the change in code that matters to replication; it is the change in substrate. Replication appears to be substrate-specific.[12] This is probably due to the fact that replication is a rather fragile process, a specialized kind of duplication which requires precise management—which means the beginning and end states must be physically similar, based in the same kind of substance. Certainly, no known replicator can replicate on more than one substrate: genes in DNA, prions as proteins, and computer viruses in computer memory. Presumably the same condition holds for memes, if they are replicators.

What is the proper substrate for memes then? It is commonly accepted that the primary repository of memes is brains. Why? Because memes are supposed to explain cultural change, and the quintessential cultural traits such as beliefs and values which distinguish one culture from another are in people's heads. Presumably for reasons like this, Dawkins, in his second book, *The Extended Phenotype*, restricted a meme to being 'a unit of information residing in the brain'.[13]

But we still have a conceptual problem: if replicators are restricted to single substrates, how can we explain processes in which replicators appear to switch substrates, as in our story about a gene above? If genes don't exist in artefacts, then how can we account for a life history in which a gene passes through a phase in which it exists only as a piece of paper? How can a second copy of DNA acquire its genetic information (in our story above) if it has had no contact with its creator, the original bit of DNA? The answer is that the gene must be reconstructed from the information that *is* present in the symbolic sequence on paper, and which bears some relationship to the gene sequence. Fancy machines must reverse engineer the gene from the information on the piece of paper.

I have argued that a similar process occurs in the case of memes: when someone reads a book, and thereby acquires the author's ideas without ever meeting the author face-to-face, the book has served as a template, holding information that creates complex visual signals which, when perceived by the reader, instigates the reconstruction of the author's memes in the reader's mind.[14] Just as the gene in the story above is reconstituted in DNA based on a paper-based representation, so too can a meme be reconstructed from a representation found on a piece of paper in a book.

Even face-to-face communication relies on the ability of human minds to engage in the reconstruction of information. This is because brains don't come into direct contact with one another. To jump the gap between minds, memes must use a signalling system, such as speech. This in turn means that message receivers must reconstruct a meme from the information contained in the signals it produces. The central question then is whether this mental reconstruction process can result in a copy of the original meme being produced—whether reconstruction satisfies the conditions for a replication process outlined above.

At present, it is difficult to know because we don't currently understand how social learning occurs. However, there are suggestions, largely from linguistics—the study of the most sophisticated natural signalling system known—that 'copying the product' (as Blackmore calls it) is a process fraught with difficulty. The information contained in a message is rarely sufficient to establish its meaning. Each instance of listening to someone else requires inferring not only the semantic content of the message, but also the intentions of the speaker, which may bear little relationship to the message. For example, ironic communication is based on saying the reverse of what you mean (e.g. 'I love your hairdo'). To make sure that the communication results in the receiver interpreting a message in something like the way the speaker intends, there have to be complex regularizing mechanisms (to eliminate spurious or extraneous elements) and a lot of shared background knowledge. So interpersonal communication is an instance of the same kind of process as learning from artefacts: a constructive process based on inadequate signals—but this time received from an active, as opposed to an inert, source.

Even the most efficient form of social learning, imitation, which is supposed to ensure high fidelity copying, is likely to introduce variation into what is learned.[15] Exact copying is not a feature to be expected of everyday human communication because the signals we send are highly impoverished compared to what we infer from them. Certainly, many of the experimental studies of cultural transmission show rapid decay in messages, and reversion to 'lowest common denominator' content.[16] If this is the case, it seems unlikely that culture can be viewed productively as the creation of lineages of information transmission with high fidelity duplication and the long-term maintenance of cultural content. Human communication systems are thus unlikely to involve replicator-like inheritance—at least in the preponderance of cases. As a result, there may be no such thing as memes, in the strict sense.

More fundamentally, communication, when seen from an evolutionary point of view, is not *designed* to result in the copying of information. Another of Dawkins' major contributions was to point out that communication is a form of signalling designed to manipulate the minds, and hence the behaviour, of other animals.[17] It is often in an individual's interest to get others to behave in ways which provide them benefits they can't achieve for themselves. This can be achieved by sending others information about a purported change in circumstance on which those individuals will then feel they should act. In some cases, what the communicator wishes won't be in the best interests of those listening, so the communicator will want to hide his or her true ambition—not only from the message receivers, but often from themselves as well, so as to more 'honestly' signal their apparent, deceptive intent.[18] From the message receiver's point of view, it will be important to make sure others are not trying to influence you in detrimental ways. Message receivers will only care about copying what is in someone else's head if that information is relevant to them, in their situation. But this won't often be the case, given that individuals are typically in different situations, with different interests.[19]

From this perspective, communication is not a peaceful exchange of information but rather tacit interpersonal warfare using information as a weapon. Of course when genetic or social interests overlap, communication can be cooperative, and information copying might be a desired outcome of a message passing between cooperators. However, most cooperation requires people to adopt complementary, rather than similar, roles. Think of a simple example: two people trying to move a piano upstairs, one going backwards, the other forwards. Here, most of the shouting is about persuading the partner to move their bit of the piano to the left or right. Even in such cases, it doesn't seem necessary to know what is in the other fellow's mind to succeed. The knowledge of each cooperator can remain quite distinct.

Of course, even if communication isn't about replicating knowledge, memes might still be able to parasitize the communication process in order to duplicate themselves. However, if the objective of communication is primarily to manipulate those with different interests to oneself, natural selection should be expected to have evolved mechanisms for persuading others, not for copying information. So it might be difficult for memes to find ways to replicate information when duplication facilities have not evolved.

There is another difficulty to mention. We have relied throughout this discussion on the commonsensical assumption that memes occupy slots in the brain for different cultural traits. For example, the meme meme is one candidate for the 'name of unit of cultural transmission' slot, competing with culturgen as well as other terms as possible values. Folk psychology suggests the existence of such a concept, but perhaps the brain doesn't work that way; perhaps it represents information rather differently than the analogy to a filing system would imply. This possibility opens whole new vistas for what memes might be in conceptual terms: not units of language like words or even abstractions like concepts (such as the meme meme), but something our conscious minds cannot conceive of—perhaps something as alien to folk psychology as the computer representation of words in binary digits.[20] The whole project of counting words on web pages or even instances of mental concepts in brains may be misguided. My own view is that memetics can only really take off once we have a better idea of how brains manage information, much like biology blossomed after the discovery of the DNA-based mechanism of gene replication. However, if it turns out that social learning typically doesn't involve the replication of information, then models of cultural evolution other than memetics will be necessary.

For a number of reasons, then, the replication of information is unlikely to be how most social learning occurs. Neither are memes necessary to explain cultural traditions. Henrich and Boyd have shown that even if copying is sloppy when individuals communicate with each other, the result of lots of sloppy social learning, when aggregated to the population level, can appear like a replication-based process in the sense that cultural traditions can still be maintained and adaptations can accumulate over time.[21] This is true if one assumes that human psychology includes a tendency to favour the acquisition of specific trait values, or what Sperber calls 'cultural attractors'. Thus, even if memes aren't at work in culture, it can appear as if they were. So taking the stability of culture as prima facie evidence of the existence of memes is mistaken. Replication is not a necessary component of an interesting Darwinian process, and may not be involved in the explanation of human culture. Dawkins presaged a similar conclusion long ago: 'My own feeling is that its main value [the meme hypothesis] may lie not so much in helping us to understand human culture as in sharpening our perspective of genetic natural selection.'[22] My attempt to provide a more precise definition of memes has, ironically, shown that memetics appears to be in search of subject matter because its central claim, the meme hypothesis, lacks substance.

A final speculation about the fate of memes: even if it turns out that there are no mental replicators, it will be difficult to deny memes a role in the future of cultural evolutionary studies. This is because the meme meme has already become part of the culture it was supposed to explain—as attested by the frequency of its mention on the Web. I therefore suspect people will continue to use the word 'meme' in a vague way when discussing cultural change. But I also predict that memetics is unlikely ever to become an empirical science, because when we define memes in a manner precise enough to start making testable predictions, we find that we have largely defined them out of existence.

The last chapter of *The Selfish Gene* has thus proven incredibly provocative, and productive—at least in the sense of having spawned renewed interest, and a burgeoning literature, in the evolution of culture. At minimum, the meme concept has shown how evolutionary biology provides a model for the study of a central concept in the social sciences: culture. Interestingly, Dawkins suggested that any process which showed design was likely to be due to the natural selection of random variants—a principle he called 'Universal Darwinism'.[23] Some have taken this idea as a rallying cry, and used the meme concept as part of a general programme to apply Darwinian principles to the disciplines bordering on biology, particularly psychology and the social sciences. This kind of theoretical unification is highly desirable, if only for the parsimonious explanations it provides for a broad range of phenomena. But of course the idea that Darwinian theory can better account for the subject matter of a discipline than theories home-grown in that discipline itself has been fiercely resisted as a form of territorial imperialism by those whose territory is being contested (e.g. Sahlins and Kitcher).[24] Nevertheless, the success of evolutionary psychology and cultural evolutionism are clear indications of the rapid spread of what might be called the 'Universal Darwinian programme'—and testimony to the fertile theoretical mind of Richard Dawkins.

ENDNOTES

1 Charles J. Lumsden and Edward O. Wilson, *Genes, Mind and Culture: The Coevolutionary Process* (Cambridge: Harvard University Press, 1981).

2 Richard Dawkins, 'Introduction', in Susan Blackmore, *The Meme Machine* (Oxford: Oxford University Press, 1999).

3 Blackmore, *The Meme Machine* (1999); Daniel C. Dennett, *Consciousness Explained* (New York: Little, Brown, 1991); Daniel C. Dennett, *Darwin's Dangerous Idea* (New York: Simon & Schuster, 1995); Robert Aunger (ed.), *Darwinizing Culture: The Status of Memetics as a Science* (Oxford: Oxford University Press, 2001); Robert Aunger, *The Electric Meme* (New York: Simon & Schuster, 2002); and Kate Distin, *The Selfish Meme: A Critical Reassessment* (Cambridge: Cambridge University Press, 2004).

4 Dawkins, 'Introduction' (1999).

5 Richard Dawkins, *The Selfish Gene* (Oxford: Oxford University Press, 1976).

6 John Bowker, *Is God a Virus?* (Oxford: SPCK, 1995).

7 Susan Blackmore, *The Meme Machine* (1999); see also 'The evolution

of meme machines', in A. Meneghetti et al. (eds.), *Ontopsychology and Memetics* (Rome: Psicological Editrice, 2003), 233–240.

8 Aunger, *The Electric Meme* (2002).

9 Aunger, *The Electric Meme* (2002).

10 David Hull and John S. Wilkins, 'Replication', *Stanford Encyclopedia of Philosophy* (http://plato.stanford.edu/entries/replication/; 2001).

11 Hull and Wilson, 'Replication' (2001).

12 Aunger, *The Electric Meme* (2002).

13 Richard Dawkins, *The Extended Phenotype* (Oxford: Oxford University Press, 1982).

14 Aunger, *The Electric Meme* (2002).

15 Dan Sperber, 'An objection to the memetic approach to culture', in Robert Aunger (ed.), *Darwinizing Culture: The Status of Memetics as a Science* (Oxford: Oxford University Press, 2000), 163–174.

16 Alex Mesoudi, *The Transmission and Evolution of Human Culture* (University of St. Andrews: Ph.D. Thesis, 2005).

17 Richard Dawkins and John Krebs, 'Animal Signals: Information or manipulation', in J. R. Krebs and N. B. Davies (eds.), *Behavioural Ecology: An Evolutionary Approach* (Oxford: Blackwell, 1978).

18 Robert Trivers, 'The elements of a scientific theory of self-deception', *Annals of the New York Academy of Sciences*, 907 (2000): 114–131.

19 Dan Sperber and Deidre Wilson, *Relevance: Communication and Cognition* (Oxford: Blackwell, 2nd edn., 1995).

20 Aunger, *The Electric Meme* (2002).

21 Joseph Henrich and Robert Boyd, 'On modeling cognition and culture: Why replicators are not necessary to cultural evolution', *Journal of Cognition and Culture*, 2 (2002): 87–112.

22 Dawkins, *The Extended Phenotype* (1982).

23 Richard Dawkins, 'Universal Darwinism', in D. S. Bendall (ed.), *Evolution from Molecules to Men* (Cambridge: Cambridge University Press, 1982), 403–425.

24 Marshall Sahlins, *Culture and Practical Reason* (Chicago: University of Chicago Press, 1976) and Philip Kitcher, *Vaulting Ambition: Sociobiology and the Quest for Human Nature* (Cambridge, MA: MIT Press, 1985).

HUMANS

SELFISH GENES AND
FAMILY RELATIONS

Martin Daly and Margo Wilson

W HAT did research on human family relations look like before Richard Dawkins made the advantages of taking a 'gene's eye view' clear and widely accessible? We'd like to say that it's hard to remember but, alas, it's all too easy, for this field is still largely pre-Dawkinsian, indeed pre-Darwinian. The result is not a pretty picture. Whereas true scientific theories are reductionistic efforts to predict and explain phenomena from more basic facts and principles, what passes for theory in family studies is often just re-description with a veneer of jargon. When mothers are seen to differ from fathers in behaviour or sentiments, for example, the differences are routinely attributed to distinct maternal and paternal 'roles', as if relabelling the observations in this way somehow explained them. One is reminded of the Monty Python sketch in which a Miss Elk (John Cleese) proclaimed her 'new theory of the brontosaurus' as follows: 'All brontosauruses are thin at one end, much thicker in the middle, and then thin again at the far end.'

Fortunately, beyond the walls of the Family Studies departments and their professional journals, there is a lively and growing body of evolution-minded theory and research on fundamental issues. The crucial missing element, which the gene's eye view provides, is an appreciation that husbands, wives, and children have some basic commonalities and conflicts of interests whose distal origins reside in the substantial but imperfect overlap of their prospects for genetic posterity ('fitness').

In human beings, as in other sexually reproducing creatures, children are the fitness 'vehicles' of both parents, and natural selection therefore promotes a commonality of purpose among mates. To a very large extent, things that affect a wife's expected fitness have parallel effects on her husband's, with the result that couples often come to see the world, with its prospects and pitfalls, through similar lenses. However, this commonality of interest is not perfect. Mates have distinct sets of kin who contribute to only one partner's 'inclusive fitness',[1] and this is surely the fundamental reason why in-laws are a cross-culturally ubiquitous source of marital conflict. Moreover, extramarital opportunities can tempt marriage partners and undermine their shared purpose, which is especially devastating to fitness for a man who is cuckolded and unwittingly invests his efforts in the rearing of other men's children. This is probably why adultery is the most emotionally charged source of marital conflict, and why men have repeatedly been found to resent it even more than women.[2]

In 1974, the American biologist Robert Trivers extended evolution-minded analysis of family relations with a simple, compelling theory of conflicts between parents and their young, and between siblings.[3] Because raising successful young is the principal route to fitness, most of us had implicitly assumed that the genetic interests of parents were essentially identical to those of their children, but Trivers explained why they are not. From a parent's perspective two children of equal quality are equally worthy of nurture, but from each child's perspective, one's self ('Ego') is more valuable than a sibling ('Sib'), because Sib carries only half of Ego's genes, if they share two parents, and only a quarter if they share but one. The upshot is that selection will favour youngsters who crave a larger share of parental nurture and are more selfish in their interactions with their siblings than would be ideal from the parent's point of view.

With these insights as inspiration, evolutionists began asking some basic questions. What exacerbates and what mitigates marital conflict, sibling conflict, parent–offspring conflict? How do we know who our relatives are anyway, and how do we respond to attributes such as facial resemblance that carry information, albeit fallible information, about shared genes? And has the fact that ancestral men were sometimes fooled about paternity whereas maternal links were never in doubt led to the evolution of sex differences in those parts of the brain/mind that respond to these indicators of kinship? Answers to these and other such questions are now rolling in, and the adaptations that they reveal can be elegant and subtle.

Facial resemblance, for example, is a candidate cue of genetic relatedness whose impacts have been studied extensively by Lisa DeBruine of Aberdeen University.[4] In her experiments, participants respond to computer-displayed photographs of what they believe to be real people, manipulated to look, ever so slightly and in a way that is not consciously detected or suspected, like the participants themselves. Such subliminal facial resemblance both enhances respondents' willingness to entrust a decision with real monetary consequences to a stranger and detracts from the pictured person's sex appeal, superficially contrary responses that DeBruine had specifically predicted. Why? Because our genetic relatives are at one and the same time appropriate targets of prosocial behaviour (because they are contributors to our inclusive fitness) and inappropriate mates (because of the genetic costs of inbreeding). No psychologist whose imagination was uninformed by Darwinism would ever think to investigate such matters.

Some other questions about family affairs are still, surprisingly, wide open to investigation. When a mother produces a second or subsequent child, for example, does her toddler adjust the intensity of sibling conflict in response to information bearing on whether the baby is likely to be a full sibling (same father) or a half sibling (new father)? This question arises when we adopt the gene's eye view because the toddler must strike a balance between self-interested demands for parental resources and concomitant threats to the baby's well-being, and the distinction between a full and a half sibling is not trivial in this calculus. For full siblings, each of Ego's genes has a 0.5 probability of being represented in Sib by a copy inherited from a common parent, but this drops to 0.25 if Sib has a new father, making Sib a much less promising contributor to Ego's eventual fitness. Moreover, ethnographic studies of hunter-gatherers living much as our ancestors must have done suggest that siblings of both types occurred with sufficient prevalence during our evolutionary history for this distinction to have exerted a strong selection pressure. Nevertheless, as far as we know, there is no published research on this question.

No published research on human beings, that is. Animal behaviourists have long been aware that there are circumstances in which differentiating between full siblings and maternal half siblings could be valuable, and in 1982, Warren Holmes and Paul Sherman showed that certain ground squirrels could and did do just that.[5] This was a particularly interesting case because the discriminations were made within groups of litter-mates. Female squirrels routinely mate with several males in rapid succession, with the result that even a womb-mate might be a full or a half sibling, and what Holmes and Sherman found was that full sib litter-mate sisters cooperated more and competed less, once they had become adults, than did half sib litter-mate sisters. Because there are unlikely to be circumstantial cues by which a female squirrel could make this distinction, the researchers inferred that she uses 'self-referential phenotype matching', that is, that she compares each sister to herself with respect to some genetically complex trait or traits (probably odours) and adjusts her responses to the sister accordingly.

Although no such cases had yet been described when Dawkins wrote his second book, *The Extended Phenotype*, he presciently discussed where we might expect to see self-referential phenotype matching of the sort that Holmes and Sherman's squirrels probably use, and he dubbed it 'the armpit effect'.[6] Only quite recently has this hypothesized process attracted much research interest,[7] and there is now some reason to suspect that it may be of relevance even in our own species. We say this because it has been reported that women react to body odours which indicate that the odoriferous party shares alleles (particular versions of genes) with the sniffer's parents in different ways depending, first, on whether they are the particular alleles that the sniffer herself did or did not inherit and, second, on whether the alleles matching her own were inherited from her mother or her father.[8] This remarkable result suggests that we may possess evolved capabilities specialized not only for kin detection, but for finer discriminations including differential assessment of matrilateral and patrilateral relatives. The more one thinks about this, the less far-fetched it seems to propose that toddlers might be in the business of assessing the paternity of their newborn siblings. Someone really must try to find out.

If your uterine siblings might find it useful to assess your paternity, think how much more interesting evidence bearing on this question must be to your putative father. Parental effort is a precious resource, and selection favours expending it in pursuits that are likely to promote the parent's fitness. From the gene's eye view, labouring to raise a rival's offspring is a disastrous mistake. The European cuckoo provides an iconic case in point, much loved by Dawkins himself and especially well illuminated by his colleague Nick Davies. Cuckoos lay their eggs in the nests of other species, leaving these duped 'hosts' to raise them, and their doing so has exerted a sufficient selection pressure on certain host species that a variety of specialized anti-cuckoo adaptations have evolved, to which the cuckoos have in turn responded, producing some classic examples of the coevolutionary process that Dawkins made vivid with the label 'evolutionary arms race'.[9]

A perceived parallel between the cuckoo's parasitic behaviour and human adultery lies at the origin of the word cuckold, which is 'a derisive name for the husband of an unfaithful wife' (Oxford English Dictionary). And why is the cuckold considered a pitiful loser? The etymology shows that the risk of mistakenly rearing another's child is central. Indeed, there is more direct evidence that people didn't have to wait for Darwin to understand that this risk underpins men's abhorrence of female infidelity (which, by the way, is found in every society yet described, recurrent claims to the contrary notwithstanding.)[10] After the French Revolution, for example, progressive lawmakers sought to abolish unjust discrimination, including that based on gender, but they uniquely exempted adultery law from such reforms, reasoning thus:

It is not adultery per se that the law punishes, but only the possible introduction of alien children into the family, and even the uncertainty that adultery creates in this regard. Adultery by the husband has no such consequences.

Paternal care is quite rare in mammals, and uncertain paternity may be why (although Dawkins defended an alternative explanation in *The Selfish Gene*).[11] Nevertheless, *Homo sapiens* is one species in which males invest substantially in the care and rearing of young, and we might therefore expect men to be sensitive to available information about paternity. In this light, it is unsurprising that there is a great deal more interest in a newborn baby's resemblance to its father than to its mother, nor that mothers seem to be especially highly motivated to detect and remark upon such resemblances.[12] A study suggesting that babies really *do* resemble their fathers more than their mothers[13] created a stir some years ago, but more thorough investigations have failed to replicate it, leaving most interested scientists convinced that the initial finding was a fluke.[14]

But would it even be in a baby's interests to advertise its paternity? If mum's consort is indeed dad, he may be pleased, and his labours co-opted, by some clear sign. But what if he is not the father? Cuckoo eggs have evolved to mimic those of their hosts in order to avoid being detected and rejected. By analogy, might babies be indiscriminable blobs *by design*? There have been several theoretical treatments of this issue,[15] but in our view, they all suffer from dubious implicit or explicit assumptions about a pre-existing, stable repertoire of paternal responses to resemblance. In a more realistic model, the baby's phenotypic expression of potential paternity cues, the father's perception of those cues, and the father's reaction to those perceptions (not to mention the responses of other interested parties) must all be permitted to coevolve. This complexity may defy theoretical analysis, at least for the time being, but some enlightenment might be obtained from computer simulations like those of Robert Axelrod and his colleagues.[16]

When we think about human families, the behaviour of birds can be a richer source of insights than the behaviour of animals that are closer kin to us.[17] The primary reason for saying this is that, as is the case in most human societies but in relatively few mammals, most young birds are raised by cooperating mated pairs. According to a famous estimate by David Lack, 92 per cent of bird species are monogamous, at least for a given season. But what none of us suspected in the 1970s, when we were just learning to think of mates as having not only a common purpose but conflicts of interest as well, was the remarkable incidence of 'extrapair paternity' (adulterous conception) in many birds, especially songbirds.

In the early 1980s, ornithologists began publishing genetic analyses showing that one species after another had cuckoldry rates of 30 per cent or more. These findings gave sociobiologists quite a shock, and complicated the task of developing good theoretical models of conflict between the sexes in ways that have not yet been put to rest.[18] Further studies showed that many species really are monogamous genetically as well as socially, and many more have extrapair paternity rates near, but not quite at, zero. How to explain the cross-species diversity in this phenomenon is still very much up in the air,[19] but what is of special interest here is that males are often devoted caretakers of the helpless young in their nests even in species in which the extrapair paternity rate is very high. In some species, males do at least modulate their caretaking efforts in response to cues indicative of their share of paternity, but in others they apparently do not. Why male birds so often soldier on to raise chicks that were sired by their neighbours is still a big puzzle. Partial answers include that males do at least target their care towards those young who have the highest statistical chance of being their own, and that there may be nothing better to do if you're a seasonal breeder and the females are no longer fertile.

We used to think that the very existence of paternal investment in *Homo sapiens* was evidence that our male ancestors could nod off at night secure in their confidence of their wives' fidelity. But when we found out that male Tree Swallows care assiduously for nestlings fewer than half of whom are their own genetic progeny, it was time to rethink the matter. Maybe not everything that looks like paternal care really functions as such. Maybe some of it is 'mating effort' instead: the male's way of trying to purchase the paternity of the female's next child. This idea seems to sit well with the distribution of step-parental care in animals generally,[20] and it may well have something to do with men's investments in their genetic offspring, too.[21]

The incidence of extrapair paternity in our own species is a topic that fascinates many, but the truth remains elusive.[22] You might suppose that modern genetic methods would have settled the matter, but they have not. Some notorious estimates on the order of 30 per cent are based on cases in which doubting fathers specifically sought the test, which hardly qualifies them as representative of the population! Certain other estimates based on blood or tissue samples collected for other purposes may also be too high, because true cuckoldry was not distinguished from cases of stepchild adoption. Changing ethical standards increasingly preclude improved replications of such analyses without the informed consent of the parties involved, so we may never find out. What must surely be true, however, is that there is no one magic number that represents 'the' human extrapair paternity rate. There is too much cross-cultural diversity in marital and sexual practices for that to be plausible.

In Chapter 6 of *The Selfish Gene*, Dawkins considered the case of societies in which men's confidence of paternity is low and suggested that they might therefore prefer to invest in uterine relatives that are surefire kin, rather than in their wives' children. This was not a new idea, and Dawkins probably got it from Alexander, a paper that discusses it extensively;[23] Dawkins apologetically acknowledges as much in the preface to the 1989 edition. There is good evidence that the hypothesis is right: the social institution called 'the avunculate', whereby men invest in their sisters' sons and make them their heirs, indeed occurs in societies with weak constraints on female sexuality and resultant uncertainty about who fathered whom.[24] There is a complication, however. A little calculation shows that the chronic societal level of extramarital paternity that would be required for sisters' sons to become closer relatives, on average, than wives' sons is 73 per cent, and there is no reason to imagine that cuckoldry rates are that high in even the most permissive societies! This is where it becomes important to think about the conflicting views of other interested parties. Whereas a man might prefer his son to his sister's son unless paternity uncertainty is very high, both potential heirs are grandsons ($r = .25$) to his parents, and they should prefer that family property go to their daughter's boy rather than their son's if there is any uncertainty whatever. So if the parents (and other relatives) exert partial influence, it is not after all surprising that avuncular inheritance should prevail below the 73 per cent threshold.[25] Careful consideration of the ways in which puzzling social institutions might represent compromise solutions to such conflicts of interest is a mental habit worth cultivating.

Richard Dawkins has often said that he never considered the gene's eye view which he expounded so effectively in 1976 a radical innovation. It had already become a familiar part of the evolutionist's toolkit, thanks largely to the contributions of W. D. Hamilton and G. C. Williams, but in Richard's view, its potential as a sort of 'universal acid' was not yet appreciated. We were working on the first draft of a book of our own in 1976, and one of us (Martin) was keeping a diary, which provides glimpses (often embarrassing) into how we then thought. It also happens to prove Richard's point. Three months prior to an entry that mentions having at last got hold of *The Selfish Gene*, Martin wrote:

Richard Alexander, on p 340 of his *Annual Review of Ecology & Systematics* paper, says 'Suppose that a juvenile mutates in such fashion as to cause an uneven distribution of parental benefits in its own favor, thereby reducing the mother's overall reproduction. A gene which in this fashion improves an individual's fitness when it is a juvenile cannot fail to lower its fitness more when it is an adult, for such mutant genes will be present in an increased proportion of the mutant individual's offspring. Thus no individual can receive a net benefit from possessing such an allele, and genetic lines will win that lack alleles disrupting in this fashion the parent-offspring interaction.'

This seemed to me so reasonable that I set about making up a simple numerical example for use in our book. And in so doing discovered it's not true!

The diary entry proceeds with details of a crude but sound model in which a mutation such as that envisioned by Alexander increases in prevalence at the expense of its 'wild-type' allele even though it damages its carriers' Darwinian fitness. Of course, Richard had scooped us: in Chapter 8 of *The Selfish Gene*, he had exposed the same theoretical error in Alexander's otherwise brilliant and wide-ranging 1974 paper. There are no other such arguments in that 1976 diary, and yet it shows clearly (and to our own surprise, thirty years later) that we actually *did* know how to make use of the perspective advocated by Dawkins' book before we read it. What we did not appreciate was the clarity that might be attained by using the selfish gene perspective unabashedly and consistently.

One of the great virtues of *The Selfish Gene* was its treatment of the human animal. Without shying away from the cultural capacity that makes our species unique, Dawkins insisted that people can usefully be thought about in the same way as other animals. This seems to us an important part of the reason why several students have told us that reading *The Selfish Gene* as high school students or undergraduates transformed their world views, and why it is still a good gift for a bright teenager. Certainly, there is still much that anthropocentric social scientists could gain from the zoological perspective that Dawkins teaches by example in all his books.

ENDNOTES

1 i.e. genetic posterity through effects on collateral as well as descendant kin; see, W. D. Hamilton, 'The genetical evolution of social behaviour', *Journal of Theoretical Biology*, 7 (1964): 1–52.

2 M. Wilson and M. Daly, 'The man who mistook his wife for a chattel', in J. H. Barkow et al. (eds.), *The Adapted Mind* (New York: Oxford University Press, 1992), 289–322.

3 R. L. Trivers, 'Parent–offspring conflict', *American Zoologist*, 14 (1974): 249–264.

4 L. M. DeBruine, 'Facial resemblance enhances trust', *Proceedings of the Royal Society of London, B*, 269 (2002): 1307–1312 and L. M. DeBruine, 'Trustworthy but not lust-worthy: Context-specific effects of facial resemblance', *Proceedings of the Royal Society of London, B*, 272 (2005): 919–922.

5 W. G. Holmes and P. W. Sherman, 'The ontogeny of kin recognition in two species of ground squirrels', *American Zoologist*, 22 (1982): 491–517.

6 R. Dawkins, *The Extended Phenotype*. (San Francisco: W. H. Freeman, 1982).

7 M. E. Hauber and P. W. Sherman, 'Self-referent phenotype matching: Theoretical considerations and empirical evidence', *Trends in Neuroscience*, 24 (2001): 607–614.

8 S. Jacob, M. K. McClintock, B. Zelano, and C. Ober, 'Paternally inherited HLA alleles are associated with women's choice of male odor', *Nature Genetics*, 30 (2002): 175–179.

9 N. B. Davies and M. Brooke, 'Coevolution of the cuckoo and its hosts', *Scientific American*, 264 (1991): 92–98; and N. E. Langmore, R. M. Kilner, S. H. M. Butchart, G. Maurer, N. B. Davies, A. Cockburn, N. A. Macgregor, A. Peters, M. J. L. Magrath, and D. K. Dowling, 'The evolution of egg rejection by cuckoo hosts in Australia and Europe', *Behavioral Ecology*, 16 (2005): 686–692.

10 M. Daly, M. Wilson, and S. J. Weghorst, 'Male sexual jealousy', *Ethology & Sociobiology*, 3 (1982): 11–27.

11 R. Dawkins, *The Selfish Gene* (Oxford: Oxford University Press, 1976).

12 M. Daly and M. Wilson, 'Whom are newborn babies said to resemble?', *Ethology & Sociobiology*, 3 (1982): 69–78; and J. M. Regalski and S. J. C. Gaulin, 'Whom are Mexican infants said to resemble? Monitoring and fostering paternal confidence in the Yucatan', *Ethology & Sociobiology*, 14 (1993): 97–113.

13 N. J. S. Christenfeld and E. A. Hill, 'Whose baby are you?', *Nature*, 378 (1995): 669.

14 P. Bressan and M. Grassi, 'Parental resemblance in one-year-olds and the Gaussian curve', *Evolution & Human Behavior*, 25 (2004): 133–141.

15 M. Pagel, 'Desperately concealing father: A theory of parent-infant resemblance', *Animal Behaviour*, 53 (1997): 973–981; R. A. Johnstone, 'Recognition and the evolution of distinctive signatures: When does it pay to reveal identity?', *Proceedings of the Royal Society of London, B*, 264 (1997): 1547–1553; and P. Bressan, 'Why babies look like their daddies: Paternity uncertainty and the evolution of self-deception in evaluating family resemblance', *Acta Ethologica*, 4 (2002): 113–118.

16 R. Axelrod, R. A. Hammond, and A. Grafen, 'Altruism via kin-selection strategies that rely on arbitrary tags with which they coevolve', *Evolution*, 58 (2004): 1833–1838.

17 S. T. Emlen, 'An evolutionary theory of the family', *Proceedings of the National Academy of Sciences, USA*, 92 (1995): 8092–8099.

18 A. I. Houston, T. Székely, and J. M. McNamara, 'Conflict between parents over care', *Trends in Ecology & Evolution*, 20 (2005): 33–38.

19 D. F. Westneat and I. R. K. Stewart, 'Extra-pair paternity in birds: causes, correlates, and conflict', *Annual Review of Ecology, Evolution & Sytematics*, 34 (2003): 365–396.

20 S. Rohwer, J. C. Herron, and M. Daly, 'Stepparental behavior as mating effort in birds and other animals', *Evolution & Human Behavior*, 20 (1999): 367–390.

21 K. G. Anderson, H. Kaplan, D. Lam, and J. Lancaster, 'Paternal care by genetic and stepfathers. II: Reports by Xhosa high school students', *Evolution & Human Behavior*, 20 (1999): 433–451.

22 K. G. Anderson, 'How well does paternity confidence match actual paternity?', under review (2005).

23 R. D. Alexander, 'The evolution of social behavior', *Annual Review of Ecology & Systematics*, 5 (1974): 325–383.

24 M. V. Flinn, 'Uterine and agnatic kinship variability', in R. D. Alexander and D. W. Tinkle (eds.), *Natural Selection and Social Behavior* (New York: Chiron, 1981), 439–475.

25 J. Hartung, 'Matrilineal inheritance: New theory and analysis', *Behavioral & Brain Sciences*, 8 (1985): 661–688.

WHY A LOT OF PEOPLE WITH SELFISH GENES ARE PRETTY NICE EXCEPT FOR THEIR HATRED OF *THE SELFISH GENE*

Randolph M. Nesse

THIRTY years ago, Western ideas about human nature bounced off *The Selfish Gene* and changed direction. Responses and related ideas continue to careen into each other with little diminished fury and successful variations are now creating their own lineages. It is a good time to assess both what *The Selfish Gene* accomplished and why so many people still hate it with such passion. The answers to these two questions are intimately related, but an analysis of the argument in *The Selfish Gene* gets nowhere without first acknowledging and seeking the source of its emotional impact.

We don't have to look far. *The Selfish Gene* illustrates, perhaps as well as any book ever written, the power of metaphor. By shamelessly anthropomorphizing genes as independent actors pursuing their selfish interests, Dawkins created wide understanding about how natural selection works that otherwise might still not exist. His use of metaphor is not only shameless, it is blameless, if you attend to the cautions he includes. Over and over again, he warns that genes are not actually actors, that they obviously are not thinking, motivated or conscious, and that the selfishness of genes is just a metaphor. These caveats slowed readers down about as effectively as 'Slow—Work Zone' signs on a deserted highway. Once his metaphor moved genes within range, our built-in capacities for intuitive social understanding snapped over them and reframed readers' minds. From the unassailable

argument that genes create organisms that act in the genes' interests, most readers followed blithely to the implication that individuals made by genes must be naturally and unavoidably selfish.
Like a surreptitious inoculation, the selfish gene metaphor slipped
a foreign idea into millions of minds where it aroused intense
reactions that sped its spread.

For me, like many others, reading *The Selfish Gene* was in equal
measures scientifically enlightening and personally disturbing.
Like most scientists in the 1970s, I had assumed that selection
shaped individuals to do what is good for the species. I thought
that helping the group was natural and this explained guilt and
other moral passions. The metaphor of the selfish gene pierced my
complacency. I saw suddenly that selection shapes actions that
advance the interests of genes no matter what the effect on groups,
species, or even individuals. Much altruism of which I was personally proud was suddenly reframed as just another way my
genes get me to do what benefits them. Selfish robots lumbered
about in my dreams for a month.

My restless nights were not mine alone. Many readers experienced the book as a psychic trauma. It turned their moral worlds
upside down. The reviews on Amazon.com include many poignant personal reports from readers, some of whom say the book
induced persisting depression. Many scientists and authors soon
began wrestling with these emotionally charged ideas. Richard
Alexander, Robert Boyd, Helena Cronin, Janet Radcliffe-Richards,
Peter Richerson, Matt Ridley, Robert Wright, and a dozen others
wrote books on evolution and cooperation.[1-7] This has now
become a flourishing research industry.[8] These intense efforts
were energized not just by curiosity, but by the moral challenge
posed by *The Selfish Gene*. Dawkins' passionate writing was, I
will wager, a response to that same moral challenge. He, like the
rest of us, was deeply disturbed by the moral implications of a
major advance in evolutionary theory.

That advance was, of course, the demise of naive group
selection. In retrospect, it is astounding that the error was not
recognized much sooner than the 1966 book *Adaptation and*

Natural Selection, by George Williams. With clear logic and vivid examples, this now classic book showed that genes for helping the group can't persist if they decrease the individual's survival and reproduction. It killed off naive group selection at a single stroke. At almost exactly the same time, William Hamilton provided the missing explanation for much helping behavior that was made otherwise mysterious by the demise of group selection. Hamilton recognized that relatives share genes that are identical by descent, so a gene that leads to helping relatives can become more common because of benefits to their children who are likely to have the same gene.

At first, these discoveries were little appreciated outside of specialized scientific circles. The 1975 publication of Edward O. Wilson's *Sociobiology* brought wide interest in evolution and animal behavior, but was not mainly about group selection and human altruism. Instead, it was *The Selfish Gene* that brought the fall of group selection and the power of kin selection to wide attention. In a display of utterly unselfish scholarship, Dawkins repeatedly gives credit to others for originating these core ideas.

Thirty years later, *The Selfish Gene* still provokes admiration, astonishment, and rage. The admiration is easily explained by the lucid prose, the astonishment by the startling ideas. But why such enduring rage? The anger arises, I think, because the main thesis of *The Selfish Gene* is not mainly about genes, it is about the behavior of individuals. The book reassesses big ancient questions about human nature in the light of the demise of group selection and gives simple unwelcome answers. Are we humans naturally good, or naturally evil? Answer: we are evil, or at least unredeemably selfish. If we are fundamentally selfish, then what explains altruism? Answer: tendencies to help others exist only if they help our genes, so helping behavior is therefore actually selfish, and true altruism is impossible or at least unnatural.

These are not abstract matters. Whether or not our attempts to help other individuals are actually altruistic or somehow covertly selfish is an emotionally charged personal issue. Almost everyone has a strong reaction. Some experience *The Selfish Gene* as a personal accusation of secret selfishness and respond with indignant rage. Others find a justification for their selfish impulses. In a book about the evolution of the capacity for commitment, I have written about twelve ways that people cope with this trauma.[9] Some try to ignore it, others attempt to show that it is false, or they attack the bearer of the news. Some try to resurrect group selection. Still others embrace it as a pure truth, long suppressed.

Like many scientists, my own habitual mechanism for coping with such traumas has been to try to figure things out. I went over Dawkins' logic again and again and couldn't see a problem. However, his conclusion didn't fit with my everyday experience, especially my work as a psychiatrist where I see so many people who spend nearly every waking minute trying to please others and feeling guilty at any hint of selfishness. To reconcile the theory and my observations, I began reading everything relevant I could find, collecting a whole shelf of books on evolution and morality, and eventually teaching a course on evolution and ethics with the moral philosopher Peter Railton. Gradually, it all worked. I finally feel I have come to grips with the challenge Dawkins posed. It has not been easy. I recommend *The Selfish Gene* to my students as a superb introduction to natural selection, but I warn them to be critical about the leap from selfish genes to selfish individuals. I hope this chapter will help them and others to take in the core message of *The Selfish Gene*, while providing some protection from undue emotional upset, and from reframing human nature as more ruthless even than it is.

I first turned to history. Sure enough, most of these ideas have bubbled over before. In 1893 T. H. Huxley published an essay on 'Evolution and Ethics', reprinted in 1989 by Williams and Paradis with their own modern commentaries.[10] What an eye-opener to find that the ethical implications of evolution have provoked consternation for over a century! And the position of my mentor George Williams is dramatic: anything shaped by natural selection is necessarily selfish so that goodness is not only not natural, it is the exact opposite of what is natural. This is very reminiscent of

Dawkins' call to 'rebel against the tyranny of the replicators'. His deductions from evolutionary theory to dark implications for human nature are in a direct line with the conclusions of some of the world's other finest thinkers.

As I ruminated about the contradiction between theory and observation, it gradually became clear that the core of *The Selfish Gene* is not a theory, a prediction, or even an observation but a logical sequence that must be true, given what we know about how selection works. Genes that make individuals with brains that give rise to behaviors that result in having more than the average number of surviving offspring will tend to become more common; individuals should, therefore, tend to behave in ways that maximize their number of offspring and reproductively successful relatives, even if those behaviors harm the group or the species. Put more succinctly, individuals are shaped to do what is best for their genes. This is incontrovertible.

What about calling such genes and behaviors 'selfish'? Genes make individuals who act to get as many of their own genes as possible into the next generation, at the expense of other individuals' genes, so that sure seems selfish. And, a gene that leads to actions that benefit others' genes more than one's own would be selected against, so such altruism seems impossible.

But pause for a moment. Are the interests of the individual really the same as those of the individual's genes? Hardly. The emotional power of the metaphor conceals the vast differences between our interests and those of our genes. This is horribly vivid in the clinic. I see scores of people who realize full well that their lusty wishes will lead to disaster, but cannot help themselves. Many others are all too aware that they have become slaves to status competition that is ruining their lives, but they persist nonetheless. Even the body's physiology reflects genes that pursue their own interests at costs to individuals, such as the shorter lifespan of males compared to females and the speed of aging. The untold story is how selfish genes give rise to emotions, behaviors, and physiological tendencies that harm the interests of the individual.

Pause again. Do our intuitions about whether an action is

altruistic or selfish depend on whether the action benefits our genes? Not at all. When a mother rushes into a burning building to rescue her child, this does not seem very selfish. Conversely, many selfish actions harm our Darwinian fitness. You don't even have to attack someone to be killed socially; in some circles, simply taking the last cookie is enough to make you a hopeless outcast. Our intuitive notions of altruism and selfishness have little to do with whether our genes benefit more than those of others. Instead, we rate actions as more altruistic in proportion to the cost of helping divided by the likelihood and amount and speed of repayment.

What about *genes* being selfish? Yes, they do everything possible to advance their own interests. But do they cheat at the expense of the whole organism? Only rarely. This is best illustrated by the few that try. Examples such as t-haplotypes in mice and segregation distorters in fruit flies manage to get themselves disproportionately represented in sperm or eggs by complex machinations often involving a pair of genes, one of which kills off cells that don't have the other half of the pair. Now *that* is nasty. It also is profoundly harmful to the individual organism and its overall reproductive success. Lawrence Hurst has even suggested that chromosomes cross over and recombine with the other paired chromosome just before creating an egg or sperm in order to separate such super-selfish gene pairs. Altruistic genes may be impossible, but cooperative ones are ubiquitous, and truly selfish genes are rare, for very good reason.

A gene gains nothing by going off selfishly on its own. Its only route to the next generation is via contributions to what Leigh has called 'The Parliament of the Genes'.[11] Genes would pursue their interests selfishly if they could, but they can't. Success comes only from cooperating with other genes to benefit the whole organism. One could write a whole book about 'The Cooperative Gene'. In an article with that title, Peter Corning notes that Dawkins is fully aware of all of this: '[Genes] collaborate and interact in inextricably complex ways, both with each other and with their external environment ... Building a leg is a multi-gene co-operative enterprise.'[12]

This cooperation is possible and necessary because all the cells in the body start off genetically identical. Muscle, bone, and skin cells have no chance of becoming eggs or sperm, so they are selected to do only what benefits the individual. Ensuring this genetic consistency is likely a major reason why life cycles reduce at one point to a single cell with a single set of genetic information and why that information is kept in a germ line sequestered from the body's other cells. Reproduction does not have to work that way. It could start with a whole cluster of cells. But it doesn't.

What about individuals? They are not genetically identical the way cells are, so they should compete to reproduce more than others in the group. They certainly do. The competition is ruthless and individuals do whatever works. But does selfish behavior work to advance the goal of maximizing reproduction? Not very often. A person who acts flagrantly selfishly even once may be ostracized for months. Conversely, a person who acts altruistically in cooperative ventures may gain huge benefits in the very long run. Game theory studies point out that altruists are subject to exploitation, but being perceived as selfish is an equal danger. Genes that make individuals who are indiscriminately selfish or generous are soon eliminated by natural selection. Like genes, individuals do whatever they can to increase the representation of their genes in future generations. Like genes, individuals accomplish this mainly by cooperating. Calling this cooperation selfish because it advances the interests of genes obscures the important differences between selfish and cooperative social strategies.

There is also an important distinction between helping that arises from calculated self-interest and helping that arises from selfless motives. We attribute much of our own helping not to calculations of how to get maximal gain, but to emotions of love, duty, and guilt. We want friends who help us out of friendship and loyalty, who do more than simply trading favors. Evidence that a supposed friend is pursuing self-interest ruins everything. If a friend gives you a ride to the airport and on the way you say, 'Well, now I owe you one ride to the airport, but only at a time of day when I am not busy and the traffic is light', your offer will never be taken up, you will never get a ride anyplace again, and your single sentence gaffe may become the subject of wide-ranging gossip. At least that is how things are here in the Midwest of the USA. The whole point of friendship is that you don't keep close score and your motives for helping are feelings, not expectations of gain. This is one reason why so many people hate an evolutionary view of human behavior. They think it implies that friendships are just exchanges, and they conclude from this that evolutionary psychologists are selfish beasts who just don't get it. The usual social response to someone who seems to be advocating selfishness is attack and social exclusion. Many authors have exercised themselves to provide such attacks and much important evolutionary science remains excluded from social sciences where it is badly needed.

Many evolutionary theorists are fully aware, however, that some human relationships involve more than kinship and reciprocity. I am particularly impressed at several comprehensive reviews of research on economic games by Ernst Fehr, each of which ends with the conclusion that we are missing something.[13] One missing concept is commitment. People make and keep commitments, sometimes even when there is no real enforcement mechanism. Furthermore, making commitments to do things that are not in your interests can be a powerful strategy of social influence. The challenge, of course, is to convince others that you will do something that is not in your interest, such as staying with and helping a spouse 'in sickness and in health'. This usually requires actually doing costly things to help others when there is no guarantee of reward. The conclusion is profound but a bit counterintuitive. People with a capacity for making and keeping commitments to do things that will not be in their interests have a strategy of social influence that gains them advantages not available to those whose behavior is predictably self-interested. These advantages are selection forces that may have shaped a capacity for commitment and moral passions to enforce them.

Such forces of social selection can shape tendencies for true altruism. By social selection I mean new forces of natural selection that emerge automatically from the dynamics of social groups.[14] This is not group selection or trait group selection or cultural group selection. It is regular natural selection at the individual level by selection forces that arise from the actions of other individuals. A simple example is the tendency to conform to social norms. The norm might be something significant such as not having sex with your cousin, or it might be just greeting others with the right hand instead of the left. Individuals who deviate from the norm are excluded. This is a potent selection force, one that I think shapes our deep human tendencies to try to figure out what others expect from us and to please them as best we can. Excessive social fears are vastly more common than lack of conscience. The complexity of human social groups gives rise to social selection that shapes human capacities for sociality different from that of other animals. Social selection seems to me to be the missing force of natural selection that explains our moral capacities, to say nothing of much interesting animal behavior. This is one of my main areas of current work.

Can natural selection really shape tendencies for true altruism as I claim above? If altruism is defined by consequences that harm the interests of one's genes, this is impossible. But selection can shape tendencies to altruistic helping that do not involve calculations or expectations of gain. True altruism provides its benefits from partnerships with others who also seek committed relationships, not exchange partners. One could try to undermine this argument with cynical reframing of such commitment as selfish. But people who believe that all others are selfish live in a social world in which that is true for them. In the clinic this is vivid. People's beliefs create social realities that repeatedly confirm the beliefs. Changing such beliefs is difficult, even if you want to and even with the help of a good therapist.

This brings us full circle to the emotional challenge posed by *The Selfish Gene*. People live by schemas based on their views of human nature, and they fight to preserve their world views, especially those close to the moral core of their identities. For many people, that makes it difficult to recognize the important truths at the center of *The Selfish Gene*. Perhaps this essay will help just a bit. If my thesis is correct, however, it won't help much.

ENDNOTES

1 R. D. Alexander, *Darwinism and Human Affairs* (Seattle: University of Washington Press, 1979).

2 R. D. Alexander, *The Biology of Moral Systems* (New York: Aldine de Gruyter, 1987).

3 R. Boyd and P. J. Richerson, *Culture and the Evolutionary Process* (Chicago: University of Chicago Press, 1985), viii, 331.

4 H. Cronin, *The Ant and the Peacock: Altruism and Sexual Selection from Darwin to Today* (New York: Cambridge University Press, 1991), xiv, 490.

5 J. Radcliffe-Richards, *Human Nature After Darwin* (Walton Hall, UK: Open University, 1999).

6 M. Ridley, *The Origins of Virtue: Human Instincts and the Evolution of Cooperation* (New York: Viking, 1st US edn., 1997), viii, 295.

7 R. Wright, *The Moral Animal: The New Science of Evolutionary Psychology* (New York: Pantheon Books, 1994).

8 P. Hammerstein, *Genetic and Cultural Evolution of Cooperation* (Cambridge, MA: MIT Press in cooperation with Dahlem University Press, 2003), xiv, 485.

9 R. M. Nesse, *Evolution and the Capacity for Commitment*, Russell Sage Foundation series on trust (New York: Russell Sage Foundation, 2001), vol. 3, page 334.

10 J. G. Paradis, T. H. Huxley, and G. C. Williams, *Evolution and Ethics: T. H. Huxley's Evolution and Ethics with New Essays on its Victorian and Sociobiological Context* (Princeton: Princeton University Press, 1989), viii, 242.

11 E. G. J. Leigh, 'How does selection reconcile individual advantage with the good of the group?, *Proceedings of the National Academy of Science USA*, 74 (1977): 4542–4546.

12 P. A. Corning, 'The Co-Operative Gene: On The Role of Synergy in Evolution', *Evolutionary Theory*, 11 (1996): 183–207.

13 E. Fehr and U. Fischbacher, 'The nature of human altruism', *Nature* 425/6960 (2003): 785–791.

14 Nesse, *Evolution and the Capacity for Commitment* (2001).

THE PERVERSE PRIMATE

Kim Sterelny

RICHARD Dawkins has had an enormous influence on my professional life. I had always been interested in the interface between philosophy and science, but until the 1980s, the sciences in question had been psychology and linguistics. (Physics was too damned hard.) But in 1983, at the urging of the philosopher Peter Godfrey-Smith, I read *The Extended Phenotype*—a truly great book—and was hooked on evolutionary biology. My copy disappeared under successive waves of marginal annotations as my career irrevocably changed course. In this essay I shall explore Darwinian visions of a somewhat neglected aspect of human nature: our paradoxical mix of astute intelligence and blindness to the obvious. I shall suggest that Richard's Darwinism, with its emphasis on conflict, helps us understand this paradox rather better than do alternative evolutionary approaches to our cognitive foibles.

Humans have invaded virtually every terrestrial habitat, and while few of us live permanently on or in water, we tax the sea's resources heavily as well. Our success has no biological precedent. While collectively the dinosaurs may have dominated land ecosystems from the Triassic to the Cretaceous, no single dinosaur species was both dominant and cosmopolitan. Likewise, while Stephen Jay Gould once argued that our time, like every time, is 'the age of the bacteria', no single bacterial species is found everywhere doing everything, even though bacteria are now and always have been by far the most numerous and ecologically varied organisms on earth. The immediate cause of our success is no mystery. It is our intelligent adaptability; our ability to assemble the technological and social means to solve the challenges the world's environments present to us. Moreover, our intelligent adaptability is an ancient rather than a modern feature of human life. One lesson anthropology teaches is an appropriate humility about the technological and ecological skills of traditional peoples. Peter Richerson's and Robert Boyd's *Not by Genes Alone*[1] richly documents these skills. To take two contrasting examples, Australian aboriginals of the desert and the Inuit of the high Arctic live in brutally unforgiving environments. Even now, it is frighteningly easy to die in these places. A friend of mine was on a geology field trip in the Pilbara during which a student died of heat exhaustion within a couple of hours of first feeling unwell and within a few hundred meters of the campsite. Arctic environments are as dangerous. Yet foragers have lived successfully in these lands for thousands of years, and they have done so without the benefits of metal technology, domestic animals, or even (in Australia) the means to store food to meet future emergencies. They survived through an intimate knowledge of their world, systems of social support, and technologies based on those materials to which they did have access.

Yet we combine this intelligence with extraordinary and destructive irrationality. We are *perversely intelligent*. A few years ago I was on a birding trip in Papua and New Guinea. Those rainforests are beautiful and diverse, but they are not well-stocked with large animals. The only large mammals are tree kangaroos, and they are rare. Protein is not supplied in large packages, and so it's no surprise that the locals have a legendarily accurate understanding of their biological environment. But they combine this understanding with deep and destructive obsessions about female menstrual pollution and about witchcraft. Many of the local cultures are tormented by fears of witchcraft and magic, and by the violence that accompanies those fears. The ethnographic record of human life documents a mix of insight and irrationality. One challenge for the evolutionary psychology of our species is that of explaining how we can be simultaneously so smart and so dumb.

Intractable idiocy would be no surprise if it were costless. But many cultures waste time, energy, and lives in protection against merely imagined danger; irrational moral, theological, social, and medical beliefs have cost thousands of lives. Nor would intractable idiocy be a surprise if we were in general cognitively incompetent, incapable of learning about subtle features of our world, and acting appropriately on the basis of that information. But we often show we have exactly that capacity. Given our capacity for competence, systematic, persistent, and expensive incompetence is puzzlingly maladaptive. How could such maladaptation persist? In an evolving population, organisms are gradually honed to suit their environments, for those that happen to fit their worlds better than their fellows leave more descendants. Since this process iterates, we expect selection to build a fit between organism and environment. *The Blind Watchmaker* and *Climbing Mount Improbable* richly document wonderful examples of adaptive design built this way. So how can an evolutionary understanding of our minds and personalities be reconciled with deep, persistent, and expensive irrationalities?

Evolutionary biologists have two main ways of explaining maladaptation. One depends on change. The kakapo—New Zealand's enormous flightless parrot—is hopeless when confronted with cats and stoats, for it evolved in a world without predatory mammals. The evolution of defences to new threats takes time, and the kakapo have not had that time. Only a few hundred years have passed since humans and their associated riff-raff first arrived in New Zealand. The environment has changed, and kakapo have yet to catch up. The other explanation depends on conflict. When a tiger stalks and kills a sambar, the sambar's adaptation for detection and flight have been trumped by the tiger's adaptations for camouflage and assault. One animal's failure is the result of the other's success.

Recent work on the evolution of the human mind—evolutionary psychology—emphasizes change as the explanation of human irrationality. One of the most striking discoveries of recent cognitive psychology is that ordinary decision problems of our daily life are informationally demanding. To make good decisions, we must be sensitive to subtle features of our environment. Think, for example, of our social worlds. Human lives are not solitary, and have not been so for hundreds of thousands of years. To live, we need to cooperate with others; we need to coordinate with others, and to coordinate without being exploited by freeloaders and shysters too severely. So we need to be able to read the intentions and emotions of others. We need to be, and are, superb *intuitive psychologists*. Poor psychologists would rarely have got to be parents of further inept judges of character, emotion, intention. Many of the typical problems of human life are demanding in similar ways, yet we respond effortlessly and successfully to most of those challenges. Steven Pinker and his allies suggest that we can do this because we have evolved a collection of special purpose cognitive machines, each of which is innately equipped to solve demanding but repeated and predictable problems of human life; he develops this view in his *How The Mind Works*.[2] As Pinker reads the human story, we are good intuitive psychologists because we have built into our minds a human psychology program—a system designed to read the thoughts and intentions of others—on which we rely as we navigate our way through the storms of our social world. We act intelligently when we face problems for which we have a well-designed module—social judgement, language, simple technology, natural history; economic exchange. We are incompetent when we face a new problem, or an old problem which has been transformed by changes to our world. Very likely, we are good intuitive psychologists only to the extent that our current world resembles our ancestral world. We are far from ept in reading the emotional tone of email messages and in judging the sincerity of mass media advertising. No wonder: Pleistocene social worlds were small and intimate. We were good intuitive psychologists in that world, and we are good intuitive psychologists to the extent that our current world resembles that ancient world in which our minds evolved.

There is something right about the idea that we are irrational about the new, but I do not think it is the whole story. We may be good intuitive psychologists, but we are appalling intuitive doctors. In the western world, a huge alternative medicine industry has been erected on the edifice of human gullibility. The folk medicines of traditional societies are typically at best ineffective; often worse. Witchcraft, shamanism, and related catastrophes are in part failed medical traditions. Why is this so? It is true that some aspects of our medical environment are new, for the little critters evolve so fast. But much is stable. Relatively simple sanitary procedures would protect against many of life's disasters. That was as true in the Pleistocene as it is now.

Richard's views on human irrationality emphasize conflict more than change, and I suspect he is right. The (apparently) irrational is sometimes a side effect of conflicts of interest. Individual rationality sometimes sums to collective idiocy. Easter Island is a famous example of humans at their most (self-)destructive: an island paradise reduced to an eroded barren wasteland, littered with broken statues and warring clans. In his book *Collapse*, Jared Diamond wonders about the psychology of over-exploitation, imagining it as self-deceptive wishful thinking:

I have often asked myself, 'What did the Easter Islander who cut down the last palm tree say while he was doing it?'. Like modern loggers, did he shout 'Jobs, not trees'? Or 'Technology will solve our problems, never fear, we'll find a substitute for wood'.[3]

Much more likely, he shouted: 'at least those bastards in the XYZ clan won't get this one'. The perspective of *The Selfish Gene* sensitizes us to a deep fact of nature: cooperative organization needs special explanation. There are circumstances in which selfish agents, agents that care only for their own benefit, cooperate. Cooperation is powerful: it generates benefits. A group of hunters can capture much larger animals than any individual acting alone can manage, and can see off dangers which would be a real threat to a solitary individual. Thus selection can favour cooperation. However, cooperation is often unstable. For a collectively produced benefit often comes with a temptation to freeload: to collect a share of that benefit while minimizing your contribution. Selection often favours successful freeloading. Crucially, selection favours freeloading even if freeloading makes everyone—even the freeloader—worse off than they would have been if no one freeloaded. For the freeloader is *relatively fitter* than good citizens, even if all are worse off than they used to be. A freeloader makes the world worse for everyone, but he makes it *especially bad* for those who are not freeloaders. And thus selection favours freeloading. The Easter Islander who pushed the Easter Island palm into extinction was relatively better off than those who had no access to that wood, and that is true even if everyone would have been advantaged by tree conservation. Much that is irrational in human worlds is the result of a conflict between what is good for a society as a whole, and what is good for specific individuals within society. Before the final ecological collapse on Easter Island, that society expended a huge chunk of its surplus in erecting huge stone statues (up to 75 tons), in a status competition between the chiefs of the competing clans. The practice imposed huge costs on most of the population. But it was very likely beneficial to the elite whose power was thus advertised. In the face of folly, we must always ask ourselves: folly for whom?

Richard has written vividly and persuasively on these issues, showing just how precarious collective cooperation is in the face of the temptation to cheat. Human groups (like animal groups) act as organized integrated collectives only under special conditions. The puzzle of human cooperation is not its occasional failure but its extraordinary persistence. But even more famous is his extra twist on conflict-based accounts of irrationality; an idea that depends on a crucial difference between genetic and cultural inheritance. Stable cooperation depends on a shared fate, when the success of one depends on the success of all. In a submarine under attack, there is no temptation to cheat. Either everyone will survive, or no one will. The crew of a submarine share a common fate. In such circumstances, even the selfish should cooperate wholeheartedly, for there is no way of harvesting the benefits of collective action without paying your share of the costs. My nuclear genes share a common fate: none can replicate unless I reproduce. As I reproduce, each of my nuclear genes have a fair and equal chance of sending a copy to the next generation. Their best option is to cooperate: to jointly build a human with the best possible prospects. My nuclear genes cooperate because they are like submariners: they have a common fate. Their route to the future is through successful reproduction. But this is not true of many of the other genes in and on me. We are ecosystems, not organisms. Each of us is home to myriads of bacteria and viruses. And many of us harbour some assortment of parasitic worms, protists, fungi, and lice. Their fate need not be tied to the fate of our nuclear genes. For they can spread horizontally and obliquely—to other members of our own generation, and to unrelated members of the next generation. It is often in the interests of (say) a virus in you to sacrifice your survivability to enhance its own transmission to new hosts. That is why viruses are sometimes virulent. It is not always so. Many animals harbour bacterial passengers that are transmitted only to their own offspring, and those passengers do not sacrifice their hosts. It would be suicide to do so.

As Richard has shown, our ideas are more like our bacteria than they are like our nuclear genes. Each of us is in a position to receive and to spread our thoughts to friends, colleagues, and acquaintances. This feature of our informational world is enormously advantageous: it enables us to take advantage of the discoveries of others without paying the costs of time, effort, and risk that made those discoveries possible. But with this advantage comes an inescapable danger. Ideas escape a filter through which our genes must pass. Those unfortunate enough to be carrying truly destructive genes cannot pass them on; those who carry somewhat unfortunate genes have difficulty in passing them on. Those who carry truly destructive ideas cannot pass them onto their children, for they do not get to have any. But they have a chance of passing them on to those they meet. Addicts, for example, can and do transmit their habits to their associates. Notoriously, in the essay 'Viruses of the Mind' reprinted in *A Devil's Chaplain*, Richard sees religion through this lens. Religion is a cognitive infection; it results from the virus-like spread of ideas. For people spread their religions, like their colds, to friends, colleagues, neighbours, and their children. Evangelical religions, on this view, are especially likely to induce self-destructive acts which increase the salience of ideas to others at the expense of their bearers' future welfare. For these religions make their bearers especially likely to spread their ideas to others. Martyrdom is the limiting case of such a display, advertising the power of religious ideas at the martyr's expense.

Richard's image is wonderfully vivid, especially for those of us who share his dim view of the intellectual and moral merits of religion. For those sceptics, there is great appeal in thinking of, say, southern Baptism as a suppurating sore rotting the neocortex of some anti-abortion demonstrator. But is this idea anything more than a vivid and provocative metaphor? For why are we vulnerable to such intellectual contagions? The cognoviral theory of religious and other irrationalities fails to explain the credibility of irrational ideas. When the London bombers embraced martyrdom, not everyone will have thought that they were utter idiots. Why are we vulnerable to evangelical religion, not to mention alternative medicine quackery or celebrities selling their eponymous knickers at inflated prices? Questions of this kind have led Dan Sperber (in his *Explaining Culture*)[4] and Pascal Boyer (in his *Religion Explained*)[5] to reject Dawkins' picture of the roots of religion. As they see it, to explain religion what we really need to know is why humans find religious ideas salient, credible, memorable. Religion would not be part of human social life if people found religious ideas absurd, offensive, or unintelligible. On their view, religion's roots lie internal to human minds. Pascal Boyer has suggested that religious ideas appeal to us because they are the right blend of the mysterious and the familiar. The denizens of religious thought are transformed familiars: creatures, but not ordinary creatures; figures human-like in their passions and emotions but not in their powers; mountains, but not ordinary mountains. The transformations make them salient; their links to the familiar make them comprehensible.

We can think of Boyer and Sperber as complementing Richard's perspective rather than contradicting it. Richard's explanation of

religion is primarily external: his cognovirus explanation emphasizes human social environments—networks of the exchange of
ideas and information. The shapes of those networks have implications for the transmission potential of human ideas, once they
have been formulated and articulated. In small-scale social
environments in which people transmit many of their ideas to
their kin in the next generation, and in which there is a reciprocal
dependence on one's neighbours, ideas and genes will be filtered
in somewhat similar ways. In large-scale and more open networks,
the transmission potential of ideas will be very different from that
of genes. But these facts about the social organization of information flow do not explain why some ideas are readily formulated
and transmitted, and others are not. All selection depends on
variation, and Dawkins' cognoviral theory of religion is not a
theory of variation. Sperber and Boyer do have a theory of variation. For this reason, we might see them as providing what we
need to turn Richard's image into a full explanation: a cognitive
bias in favour of the transformed familiar + horizontal idea flow
= cognitive disaster.

I doubt that Sperber or Boyer would agree to this reconciliation,
and in this they mirror an earlier debate between Stephen Jay
Gould and Dawkins; a debate about the history of life as a whole.
That debate was about the relative importance of selection and
the supply of variation. Gould believed that crucial features about
the history of life are explained by biases in the supply of variation rather than selection. As he saw it, for the past 500 million
years animal evolution has been surprisingly conservative, conservative because radical variants on existing animals no longer
appear. Without variation, there is no selection. So constraints on
variation and that alone explain conservatism. The jury is still out
on Gould's ideas. No one knows the extent to which great patterns in the history of life are determined by limits and biases in
variation. Likewise, the jury is still out on Sperber's and Boyer's
ideas. No one knows the extent to which innate structures of the
human mind limit the range of ideas we can formulate and spread,
for good and ill. That said, contemporary cognitive science tends

to support Sperber and Boyer, emphasizing the extent to which our minds are pre-equipped with the information that they will need. Gary Marcus reviews these ideas in his *The Birth of the Mind*.[6] I remain unconvinced. Perhaps humans are irrational in similar ways in every culture, independently of the variations in patterns of information flow across cultures. That would leave our original puzzle unsolved. Superstitious beliefs are expensive, so we would expect effective selection against the cognitive biases that produce them. Moreover, I doubt that those theories that emphasize the pre-equipment of the mind really take into account the variety of human experience and environment at and across time.

Selection can pre-equip our minds with the information they need only if those needs remain the same over the generations. I think we have been selected to cope with unpredictable circumstances, rather than being adapted to a specific ancient world. Even ancient humans lived too differently from one another for that. There never was a single ancient environment to which our minds are adapted. So for what it is worth, I bet against Sperber and Boyer; against the hunch that there are strong innate constraints on what we can think, come to believe, and persuade others of.

We cannot escape irrationality, for it's a price we pay for relying so heavily on the views of others. Our ideas (I conjecture) are formulated in response to contingent and variable aspects of our physical and social environment, so we cannot be hard-wired to strangle our hopeless thoughts at their birth. Those on whom we rely for information are often not kin, so the ideas we take up are not filtered by their effects on the life prospects of their carriers. As those ideas are often about the elsewhere and the elsewhen (that is what makes them useful) we often cannot check their truth directly. We do have indirect checks—we can evaluate the coherence of what we are told, and the reliability of the teller— but while no doubt these indirect checks protect us from much nonsense, they are far from perfect. Like many of our diseases, pathological thought is part of the price we pay for large-scale cultural life.

ENDNOTES

1   Peter J. Richerson and Robert Boyd, *Not by Genes Alone* (Chicago: University of Chicago Press, 2004).

2   Steven Pinker, *How the Mind Works* (New York: W. W. Norton, 1997; London: Penguin, 1998).

3   Jared Diamond, *Collapse* (London: Penguin, 2004), 114.

4   Dan Sperber, *Explaining Culture* (Oxford: Blackwell, 1996).

5   Pascal Boyer, *Religion Explained* (Oxford: William Heinemann, 2001).

6   Gary Marcus, *The Birth of the Mind* (Basic Books, 2004).

CONTROVERSY

THE SKEPTIC'S CHAPLAIN: RICHARD DAWKINS AS A FOUNTAINHEAD OF SKEPTICISM

Michael Shermer

OVER the weekend of 12 to 14 August 2001, I participated in an event entitled 'Humanity 3000', whose mission it was to bring together 'prominent thinkers from around the world in a multidisciplinary framework to ponder issues that are most likely to have a significant impact on the long-term future of humanity'. Sponsored by the Foundation for the Future—a non-profit think tank in Seattle founded by aerospace engineer and benefactor Walter P. Kistler—long-term is defined as a millennium. We were tasked with the job of prognosticating what the world will be like in the year 3000.

Yeah, sure. As Yogi Berra said, 'It's tough to make predictions, especially about the future'. If such a workshop were held in 1950 would anyone have anticipated the World Wide Web? If we cannot prognosticate fifty years in the future, what chance do we have of saying anything significant about a date twenty times more distant? And please note the date of this conference—needless to say, not one of us realized that we were a month away from the event that would redefine the modern world with a date that will live in infamy. It was a fool's invitation, which I accepted with relish. Who could resist sitting around a room talking about the most interesting questions of our time, and possibly future times, with a bunch of really smart and interesting people. To name but a few with whom I shared beliefs and beer: science writer Ronald Bailey, environmentalist Connie Barlow, twins expert Thomas Bouchard, neuroscientist William Calvin, educational psychologist Arthur Jensen, mathematician and critic Norman Levitt, memory expert Elizabeth Loftus, evolutionary biologist Edward O. Wilson, and many others, all highly regarded in their fields, well published, often controversial, and always relevant.

Also in attendance, there to receive the $100,000 Kistler Prize 'for original work that investigates the social implications of genetics', was the Oxford University evolutionary biologist Richard Dawkins. (Ed Wilson was the previous year's winner and was there to co-present the award, along with Walter Kistler, to Richard.) Dawkins was awarded a gold medal and a check for his work 'that redirected the focus of the "levels of selection" debate away from the individual animal as the unit of evolution to the genes, and what he has called their extended phenotypes'. Simultaneously, the award description continues, Dawkins 'applied a Darwinian view to culture through the concept of memes as replicators of culture'. Finally, 'Dr Dawkins' contribution to a new understanding of the relationship between the human genome and society is that both the gene and the meme are replicators that mutate and compete in parallel and interacting struggles for their own propagation'. The prize ceremony was followed by a brilliant acceptance speech by Richard, who never fails to deliver in his role as a public intellectual (the number *one* public intellectual in England, according to *Prospect* magazine) and spokesperson for the public understanding of science.

This is not what most impressed me about Richard, however, since any professional would be expected to shine in a public forum, especially with a six-figure motivator hanging around his neck. It was during the two full days of round-table discussions, breakout sessions, fishbowl debates, and (most interestingly) coffee-break chats, where Richard stood out head-and-shoulders above this august crowd. Despite his reputation as a tough-minded egotist, Richard is, in fact, somewhat shy and quiet, a man who listens carefully, thinks through what he wants to say, and then says it with an economy of words that is a model for any would-be opinion editorialist. In one session, for example, we were to debate 'Conscious Evolution—Fantasy or Fact?' After about twenty minutes of discussion of a topic none of us had carefully defined, Richard spoke up:

I wanted to listen around the table to see if I could make out what conscious evolution is. I still haven't. It seems to be a mix-up of two, or three, or four very different things. There is the evolution of consciousness; there is what Julian Huxley would have called 'consciousness of evolution' or, the way he put it was 'man is evolution become conscious of itself.' But entirely separate from that . . . forget about consciousness and just talk about deliberate control of evolution, and then we bifurcate again into two entirely different kinds of evolution. That is genetic evolution and cultural evolution. I am not going to utter the 'm' word [memes]; everybody else keeps saying it and then looking at me, and I am going to duck out of that. I used not to think this, but I am increasingly thinking that nothing but confusion arises from confounding genetic evolution with cultural evolution, unless you are very careful about what you are doing and don't talk as though they are somehow just different aspects of the same phenomenon. Or, if they *are* different aspects of the same phenomenon, then let's hear a good case for regarding them as such.

The first response was from the futurist Michael Marien, who said: 'I would like to start back at the point where Richard Dawkins honestly said he had never heard the term *conscious evolution*. Sometimes a statement of ignorance can be very illuminating.' Indeed it can be, and Richard's candid comments throughout the weekend illuminated the conference like no one else's.

Outside of additional specifics of what Richard said, here is my overall impression of that weekend in Seattle, an observation with broader implications for Dawkins' impact on science and culture: a discussion would ensue over some issue, such as 'the factors most critical to the long-term future of humanity', and most of us panelists would jump in with our opinions, banter some particular theme back and forth for a while, then leap to another topic, hammer that into the faux-mahogany table, and so forth round and round. Richard would sit there listening, processing the verbosities of us long-winded opinionators, select his moment to lean forward and make a short observational remark or inductive inference, then sit back and collect more data. It was what happened after Richard spoke that I came to realize this is a man on a different plane, above even these stellar minds. The conversation changed, bifurcated in a new direction, with references to its source. 'You know, Richard has a point . . .', 'I'd like to comment on Richard's observation . . .', 'Going back to what Professor Dawkins said . . .' And so on. Richard Dawkins changed the conversation. He has been changing the conversation ever since 1976, when his book *The Selfish Gene* changed the way we look at ourselves and our world.

*   *   *

Humans are a hierarchical social primate species who, despite centuries of democratic rule, still long to sort themselves into pecking orders within families, schools, peer groups, social clubs, corporations, and societies. We can't help it. It is in our nature, courtesy of natural selection operating in the social sphere. As an intellectual social movement in which I am intimately involved, skepticism is subject to the same hierarchical social forces. As such, we scientists and skeptics look up to and model ourselves after our alpha leaders. In my own intellectual development there have been several who served me well in this capacity, including Carl Sagan, Stephen Jay Gould, and Richard Dawkins. They are, in fact, candles in the darkness of our demon-haunted world (in Carl's apt phrase from his skeptical manifesto). Lamentably, we lost Carl and Steve too early. How I long for one more poetic narrative on our pale blue dot in the vast cosmos, or one more elegant essay on life's complexity and history's contingency.

But thank the fates and his hearty DNA that we have Richard, who stands as a beacon of scientific skepticism and a hero to skeptics around the world. Dawkins' work has touched the skeptical movement in three areas of common concern: pseudoscience, creationism, and religion.

Dawkins' primary work on pseudoscience is *Unweaving the Rainbow*, a collection of essays centered on 'Science, Delusion and the Appetite for Wonder' (the book's subtitle). Here we see almost no limits to the breadth of Richard's interests, as he skeptically analyzes astrology, coincidences, conjurors, eyewitness accounts, fairies, flying saucers, the Gaia hypothesis, gambling fallacies, hallucinations, horoscopes, illusion, imagination, intuition, miracles, mysticism, paranormalism, post-modernism, psychic phenomena, reincarnation, Scientology, superstition, telepathy, and even *The X-Files*. Richard's analysis of these and other delusions is not debunking as such, but more positively directed toward helping us better grasp what science is by looking at what science isn't; and how we can recognize good science by seeing bad science for what it is. Deeper still is the take-home message embodied in the book's title from Keats, 'who believed that Newton had destroyed all the poetry of the rainbow by reducing it to the prismatic colours. Keats could hardly have been more wrong'. In his stead Richard offers us this insight: 'I believe that an orderly universe, one indifferent to human preoccupations, in which everything has an explanation even if we still have a long way to go before we find it, is a more beautiful, more wonderful place than a universe tricked out with capricious, *ad hoc* magic.'

Creationism is a form of pseudoscience, and the connection here is an obvious one for an evolutionary biologist who holds the job title of 'Professor of the Public Understanding of Science'. In America, at least, there is no better example of a public *mis*understanding of science than creationism, and Richard has written broadly and deeply on the subject, and he minces no words (and it is with creationists especially that Richard does not 'suffer fools gladly', as he has been accused). After the May 2005 hearings in Kansas on the proposed introduction of 'Intelligent Design' into the public school science curriculum, Dawkins fired off an opinion editorial in *The Times* (London) on 21 May, entitled 'Creationism: God's Gift to the Ignorant', that included this poignant observation that cut through yards of creationist verbiage with clarity and wit:

The standard methodology of creationists is to find some phenomenon in nature that Darwinism cannot readily explain. Darwin said: 'If it could be demonstrated that any complex organ existed which could not possibly have been formed by numerous, successive, slight modifications, my theory would absolutely break down.' Creationists mine ignorance and uncertainty in order to abuse his challenge. 'Bet you can't tell me how the elbow joint of the lesser spotted weasel frog evolved by slow gradual degrees?' If the scientist fails to give an immediate and comprehensive answer, a default conclusion is drawn: 'Right, then, the alternative theory, "intelligent design", wins by default.'

At least three of Richard's books—*The Blind Watchmaker*, *Climbing Mount Improbable*, and *River Out of Eden*—are direct challenges to creationists' arguments, although presented not as straight debunking works, but as science-advancing treatises on evolutionary theory. And Richard's latest book, *The Ancestor's Tale*, is one long answer to creationists' demand to 'show me just one transitional fossil'. Dawkins traces innumerable transitional forms, as well as numerous common ancestors, or what he calls 'concestors'—the last common ancestor shared by humans and more and more distant groups of living species—back through four billion years to the origin of heredity and the emergence of evolution. No one concestor proves that evolution happened, but together they reveal a majestic story of process over time. Richard is the Geoffrey Chaucer of life's history, and our most articulate public defender of evolution.

Creationism, of course, is nothing more than thinly disguised religion masquerading as science, as an end-run around the US Constitution's First Amendment prohibition on the government establishment of religion. Richard's views on religion, particularly when it intersects with science, are so public and controversial that they have even inspired a book by the Oxford University professor of historical theology, Alister McGrath, *Dawkins' God*,[1] a book I reviewed in *Science*.[2] The connection between science and religion for Dawkins runs like this: before Darwin, the default explanation for the apparent design found in nature was a top-down designer—God. The eighteenth-century English theologian, William Paley, formulated this into the infamous watchmaker argument: if one stumbled upon a watch on a heath, one would not assume it had always been there, as one might with a stone. A watch implies a watchmaker. Design implies a designer. Darwin provided a scientific explanation of design from the bottom up: natural selection. Since then, arguably no one has done more to make the case for bottom-up design than Dawkins, particularly in *The Blind Watchmaker*, a direct challenge to Paley. But if design comes naturally from the bottom up and not supernaturally from the top down, what place, then, for God?

Although most scientists avoid the question altogether, or take a conciliatory stance along the lines of Stephen Jay Gould's non-overlapping magisteria (NOMA), Dawkins unequivocally states in *The Blind Watchmaker*: 'Darwin made it possible to be an intellectually fulfilled atheist.' And in *River Out of Eden*: 'The universe we observe has precisely the properties we should expect if there is, at bottom, no design, no purpose, no evil and no good, nothing but blind pitiless indifference.'

Herein lies the crux of the issue, and Dawkins brooks no theological obfuscation. For example, after debunking all the quasi-scientific and pseudoscientific arguments allegedly proving God's existence, scientists are told by theologians like Alister McGrath that we should come to know God through faith. But what does that mean, exactly? In *The Selfish Gene*, Dawkins wrote that faith 'means blind trust, in the absence of evidence, even in the teeth of evidence'. This, says McGrath, 'bears little relation to any religious (or any other) sense of the word'. In its stead McGrath presents the definition of faith by the Anglican theologian W. H. Griffith-Thomas: 'It commences with the conviction of the mind based on adequate evidence; it continues in the confidence of the heart or emotions based on conviction, and it is crowned in the consent of the will, by means of which the conviction and confidence are expressed in conduct.' Such a definition—which McGrath describes as 'typical of any Christian writer'—is what Dawkins, in reference to French postmodernists, calls 'continental obscurantism'. Most of it describes the psychology of belief. The only clause of relevance to a scientist is 'adequate evidence', which raises the follow-up question, 'Is there?' Dawkins' answer is an unequivocal 'No'.

* * *

Does a scientific and evolutionary world view such as that proffered by Richard Dawkins obviate a sense of spirituality? I think not. If we define spirituality as a sense of awe and wonder about the grandeur of life and the cosmos, then science has much to offer. As proof, I shall close with a final story about Richard, and a moment we shared inside the dome of the 100-inch telescope atop Mount Wilson in Southern California. It was in this very dome, on 6 October 1923, that Edwin Hubble first realized that the fuzzy patches he was observing were not 'nebula' *within* the Milky Way galaxy, but were separate galaxies, and that the universe is bigger than anyone imagined, a *lot* bigger. Hubble subsequently discovered through this same telescope that those galaxies are all redshifted—their light is receding from us and thus stretched toward the red end of the electromagnetic spectrum—meaning that all galaxies are expanding away from one another, the result of a spectacular explosion that marked the birth of the universe. It was the first empirical data indicating that the universe had a beginning, and thus was not eternal. What could be more awe-inspiring—more numinous, magical, spiritual—than this cosmic visage of deep time and deep space?

Since I live in Altadena, on the edge of a cliff in the foothills of the San Gabriel mountains atop which Mount Wilson rests, I have had many occasions to make the trek to the telescopes. In November 2004 I arranged for a visit to the observatory for Richard, who was in town on a book tour for *The Ancestor's Tale*. As we were standing beneath the magnificent dome housing the 100-inch telescope, and reflecting on how marvelous, even miraculous, this scientist vision of the cosmos and our place in it all seemed, Richard turned to me and said, 'All of this makes me so proud of our species that it almost brings me to tears.'

I would echo the same sentiment about the works and words of Richard Dawkins.

ENDNOTES

1   Alister McGrath, *Dawkins' God* (Oxford: Blackwell, 2004).
2   Michael Shermer, 'Book Review: Alister McGrath's *Dawkins' God*', *Science* (8 April 2005): 205–206.

A FELLOW HUMANIST

Richard Harries

RICHARD Dawkins is a brilliant writer and speaker on science. His grasp of the subject and his use of vivid analogies can explain scientific concepts and make them clear even for the non-scientist. I think, for example, of his recent discussion on Radio four with Jonathan Miller on extended phenotypes. Of many examples in his writings I refer to his explanation of 'intermediates' and why creationists are confused in calling for evidence that, by definition, could never be produced in the form they seek.[1] Then there is his sense of excitement at what has been or might be intellectually explained whilst his passion for scientific truth can arouse interest in even the most unscientific mind. If pupils at school were taught science by teachers with some of these gifts, the role and status of science in our society today would be very different.

Like Richard, I am at once puzzled by and hostile towards the apparently growing influence of creationism in this country coming over from the United States. I have been very pleased to be associated with him in writing letters and articles opposing the reported teaching of creationism in at least one new school. Like him I believe that science has a proper integrity which needs to be fought for and preserved. This means that letting evidence decide, allowing evidence to modify or refute even one's most cherished notion, is fundamental; as is the scientific method of rigorous testing of hypotheses by experiment. I also have other reasons for being antagonistic to creationism. It involves an unhistorical, uncritical approach to the biblical texts. It misunderstands what those texts set out to do and as a result they belittle God and bring Christianity into disrepute.

Richard's reputation as a gifted communicator of science is

allied to that of a fierce polemicist against religion. A book of his essays is entitled *A Devil's Chaplain*.[2] I can certainly understand many reasons why people are hostile to organized religion. One which Christians are strangely reluctant to acknowledge is the moral protest, not only against certain practices of the Church in the past, but against some fundamental notions of God. Alec Vidler in his history of the Church in the nineteenth century[3] suggests that the great agnostics who turned away from the Church did so not because of the rise of science or the advent of biblical criticism, but because what the Church called upon them to believe with a sense of its own moral superiority struck them as morally inferior to their own highest beliefs and standards. I have tried to explore something of this moral protest in a modern context in my book *God Outside the Box: Why Spiritual People Object to Christianity*.[4] The great Oxford thinker Austin Farrer, arguing once that faith is natural to humanity, went on to ask why it is then that people become atheists and then continued, 'Heaven knows there is no lack of explanations. The corruptions of faith herself are so many and so appalling as to allow atheism to pass for illumination.'[5] I can also understand why people are atheists on the grounds that the presence of so much evil and human anguish in life is incompatible with belief in a loving creator. Whatever the intellectual arguments put forward, and there are many, these can never foreclose the issue and even people of profound faith will continue to be troubled by the sombre aspects of existence. The Christian faith of Darwin drained away, though he probably never completely lost it, not because of the theory of evolution as such but because of the competition and suffering in the animal world, 'Nature red in tooth and claw'.

What is more difficult to understand is why a passionate love of science should, of itself, lead to a passionate dislike of religion. Historians of science note how relatively quickly the thinking Christian public in fact accepted the theory of evolution. Frederick Temple, the then headmaster of Rugby and later Archbishop of Canterbury, preached at the service for the British Association meeting before the University of Oxford on 1 July 1860. He

argued that evolution is entirely congruous with the divine purpose which is shown 'In the slow working of natural causes'.
Developing this theme, one made explicit by Charles Kingsley,
Austin Farrer wrote, 'God not only makes the world, he makes it
make itself; or rather, he causes its innumerable constituents to
make it.'[6] The encounter of Samuel Wilberforce, my predecessor
as Bishop of Oxford, and Thomas Huxley, at that Royal Association meeting, has achieved mythic status. Serious and detailed
examination of what actually occurred has been undertaken by a
number of scholars in recent years. It appears that Huxley did
indeed win the encounter, in the minds of those who heard it. But
this was partly because the alleged remark of Wilberforce was
regarded as ungentlemanly and because he was unpopular in
Oxford. The main scientific argument was in fact put forward not
by Huxley but by Joseph Hooker. The mythic status of the meeting began to arise much later in the century, pushed by Huxley
who was anticlerical and strangely reluctant to admit that many
clergy were, in fact, adherents of Darwin's view, even when faced
with one of them who was. The quarrel between religion and
science came about not because of what Wilberforce said but
because it was what Huxley wanted.[7]

The interaction of science and religion has been a particularly
fruitful field over the last forty years or so and opinion polls all
show that the percentage of believers and unbelievers amongst
scientists is about the same as in the population as a whole. So
I am inclined to tease Richard by saying, 'Richard, there are
enough good arguments against religion without keeping on
dragging science into it.' I have in honesty to say, even in a book
devoted to his work, and in an essay which is fundamentally
admiring of Richard's writing, that I simply don't think the
arguments he puts forward against a religious view of life based
on his scientific work stand up as, I believe, Alister McGrath has
decisively shown in some detail recently.[8]

One of the aspects of Richard's writing that I want to
emphasize is his sense of awed wonder before the universe. This
obviously underlies everything that he writes and speaks but is

stated most explicitly and explored most fully in *Unweaving the Rainbow*. Richard takes issue with those who believe that a scientific view of life like his own means that everything looks bleak and meaningless. He also takes issue with those who channel their sense of wonder into the paranormal or other forms of superstition. He suggests instead that wonder arises out of the very fact of being alive in this amazing universe; that the more we know about this universe, its age, size, and complexity, the more wonderful it seems; and the inevitable fact of death and our own transience in fact enhances, rather than undermines, a sense of gratitude and awe for being here at all. Furthermore, whereas for some minds in the past, such as Blake's, the more you analyse things in their constituent parts, the more you drain away their sense of beauty, Richard argues precisely the opposite.

The mystic is content to bask in the wonder and revel in a mystery that we were not 'meant' to understand. The scientist feels the same wonder but is restless, not content; recognising the mystery is profound then adds, 'but we're working on it' . . . Our language must try to enlighten and explain, and if we fail to convey our meaning by one approach we should go to work on another. But, without losing lucidity, indeed with added lucidity, we need to reclaim for real science that style of awed wonder that moved mystics like Blake. Real science has a just entitlement to the tingle in the spine which, at a lower level, attracts the fans of 'Star Trek' and 'Doctor Who' and which, at the lowest level of all, has been lucratively high-jacked by astrologists, clairvoyants and television psychics.[9]

In short, though science might unweave the rainbow, in the sense of helping us to understand why we see the colours of the spectrum in it, we can, in the words of James Thompson's poem to Sir Isaac Newton say that the setting suns and shifting clouds declare 'How just, how beauteous the refractive law'. So Richard, who quotes the great Indian astrophysicist Subrahmanyan Chandrasekhar:

This 'shuddering before the beautiful', this incredible fact that a discovery motivated by a search after the beautiful in mathematics should find its exact replica in nature, persuades me to say that beauty is that to which the human mind responds at its deepest and most profound.[10]

A theologian will want to see this phenomenon as grounded in a reality that lies beyond the visible universe. The fact that mathematicians look for and discover equations of extraordinary elegance and beauty, and that these enable scientists to explore the true nature of physical reality, seems to cry out for an explanation. The religious explanation is that the human mind and the way the universe reveals its secrets to rational exploration is grounded in the logos, the divine rationality and ordering of all things. There is no finally compelling logical proof that can take this step for us. Nor, on the other hand, is there any finally compelling philosophical or scientific reason why that step should not be taken. But whether it is taken or not I agree entirely with Richard that a sense of awed wonder is the most appropriate response to living in the kind of universe we do, and that science, properly understood, should enhance and strengthen that sense, not undermine or drain it.

There are two further positive aspects of Richard's thought I particularly wish to emphasize. First, his conviction, derived from his scientific work, that there is objective truth to be discovered. This makes him highly critical of postmodernism and those who emphasize the cultural relativity of all views, those who say that nothing can be known for certain. They argue that all we have are shifting perspectives reflecting the changing conditions and periods in history. Richard will have none of it. Some things can be known to be true. Although there is some truth in postmodernism, which can be assimilated without selling out to a total relativity, and this has been done by the most profound thinkers such as Rowan Williams, the present Archbishop of Canterbury, Richard's protest is an important one. And although science itself is not totally immune to cultural shifts, its conviction that real knowledge can be acquired is an important corrective to some of the thinking of certain literary critics. In this, Richard is an ally of religious people.

No less important is his stress on the fact that as humans we are moral beings, capable of distinguishing right from wrong and able to transcend our narrow egoism in acts of generosity and magnanimity. We are not simply at the mercy of evolutionary forces. It is tempting for some people who immerse themselves in the details of evolution, its directionlessness, its competitive nature and 'selfish' drives (though there are also elements of cooperation in nature), to think that we cannot help but live out such a pattern. Richard stresses that we can. We can and must go against the grain of bloody competitive struggle. He himself is committed to a number of very worthwhile concerns. Though I believe that our capacity to recognize and respond to the good is, like the order and beauty of the universe, ultimately grounded in a reality beyond it, I think it is important that all of us, whether or not we are religious believers, have a profound sense of what it is to be a human being, capable of living humanely. So I'm very happy indeed to salute Richard as a fellow humanist, one who believes in the importance, dignity, and utter worthwhileness of being a human being and of trying to live humanely.

ENDNOTES

1 Richard Dawkins, *The Ancestor's Tale* (London: Weidenfeld & Nicolson, 2004), 252–261.

2 Richard Dawkins, *A Devil's Chaplain* (London: Weidenfeld & Nicolson, 2003).

3 Alec Vidler, *The Church in an Age of Revolution* (London: Penguin, 1961), 113.

4 Richard Harries, *God Outside the Box: Why Spiritual People Object to Christianity* (SPCK, 2003).

5 Austin Farrer, *Saving Belief* (London: Hodder & Stoughton, 1964), 25–26. In fairness, Farrer goes on to say, 'But the aversion from faith need not be motivated by faith's corruptions. Men turn from faith, because to acknowledge God is to acknowledge *my God*, and men either hate, or fear to admit that they have a God, or that there is any will sovereign over their own.'

6 Farrer, *Saving Belief* (1964), 51.

7 Richard Harries, 'The Encounter Between Samuel Wilberforce and Thomas Huxley', in *Modern Believing*, vol. 47, no. 1 (January 2006): 22–27.

8 Alister McGrath, *Dawkins' God* (Oxford: Blackwell, 2005).

9 Richard Dawkins, *Unweaving the Rainbow* (London: Penguin, 1999), 17–18.

10 Dawkins, *Unweaving the Rainbow* (1999), 63.

DAWKINS AND THE
VIRUS OF FAITH

A. C. Grayling

RICHARD Dawkins is surely the most frequently-cited champion or demon, depending upon point of view, in the science versus religion wars. His detractors portray him as a humourless, soulless reductionist, a threat not just to faith but to all beauty and brightness. His admirers applaud the devastating blows he lands on the votaries of superstition and unreason. I am with the applauders; and the lucidity, clarity, and force of his contributions are a major part of what makes me so.

Having encountered a Dawkins in his writings who is eloquently and indeed poetically alert to the astonishing and beautiful world revealed by science—so much more wonderful than anything the limited imaginings of religion offer—it always used to surprise me when Dawkins' detractors charged him with sourness. But then we appeared together among the guests on a television programme hosted by Melvyn Bragg, and I could see why some might get that impression. Unsmiling and distant, perhaps because he had not met the other participants before, he addressed the topic of the programme with his customary acuity, but neither before nor afterwards interacted much with the production team or fellow panellists.

As Dawkins himself would rightly say, one cannot generalize from a sample of one: a single encounter with someone on that occasion disinclined to socialize is no basis for personal judgments. But I suppose that if others have had a similar experience of him, and have generalized incorrectly from it, they have failed to consider the possibility that as an *ex cathedra* spokesman for science (that being what his chair at Oxford is endowed for), he must find it deeply tiresome to have to meet and contest, repeatedly and with equivocal success, the weight of the majority outlook in this world, which as regards the relative merits of science and religion is stubbornly ignorant, superstitious, impermeable to rational argument, lazy, narrow, shallow, and prejudiced. Dawkins is paid a salary for, in effect, rolling Sisyphus' rock up the mountain of human unreason, and his efforts to educate the world out of its stone-age mindset must often seem to him to generate far more opposition than illumination.

The case Dawkins makes against religion, and his penetrating analysis of how the virus of faith replicates itself, are among the most important of his contributions to debate in the philosophy and sociology of science—especially the latter, for it provides a sound basis for such matters as public policy arguments against 'faith-based schooling' and the teaching of religion in schools as if it were on a par with history and natural science.

Apologists for religion (in the Western world at least) have taken to defending their outlook on supposedly scientific grounds, by arguing that a deity is required to explain why the world exists—and more particularly why life exists—and that the world manifests 'intelligent design'. This is especially true in the United States, as the 'intelligent design' example shows. This is a stratagem for circumventing legal protections for secularism in US public institutions, and particularly schools, which forbid the explicit teaching of Genesis-citing 'creationism'. In Britain the usual lazy fudge persists of regarding religion and science as mutually consistent because addressed to different spheres of concern—or, even more lazily still, of regarding God as the inventor of the laws of physics and biology which, give or take the odd miracle that locally suspends or reverses them, thereafter run automatically according to plan. Finding something to agree with Stephen Jay Gould about at last, Dawkins refuses to engage with 'intelligent design' arguments on the grounds that doing so gives them unmerited publicity. But he is not reticent about the other species of claim, to the effect that a God is required to explain the existence of the universe and life.

Familiarly, such views evaporate like dew in the sunlight of science. If religion represents itself as providing explanations that compete with natural science, it has to offer testable predictions and be responsible to evidence just as science does; and on that count it signally fails. It fails also on the test of simplicity: to explain the world it has to invoke something even more mysterious and complicated than the world itself, viz. a deity—and then declares itself incompetent to explain the deity. It thus invites a regress which it arbitrarily terminates at the first step. Saying that a deity created the world is like saying that Fred made the rain fall yesterday—with no explanation of who or what Fred is.

Moreover, although you can if you wish say that your car runs on psychokinetic energy—this is Dawkins' example—it is simpler and more powerful to say that since the fuel looks, smells, and behaves like petrol, it runs on petrol. Simplicity in this focal sense is part of the explanatory and predictive power of good theories, which are those tested by observation and experiment, and are based on principles governing our understanding of more complex phenomena on the basis of their less complex parts.

A universe created and operated by a deity would, Dawkins points out, be very different from the natural universe we occupy. Some are motivated to nominate one or other of the fundamentals of the natural universe as 'God'—superstrings or Planck's constant, say—on the grounds that they are 'mysteries' which we do not (yet) understand. But an arbitrary bit of nomenclature does nothing to connect superstrings or Planck's constant to a supposed deity of the traditional, stone-age-originated kind, who forgives sins, listens to prayers, punishes people for not observing the Sabbath—or, Dawkins laconically remarks, cares about 'whether you wear a veil or have a bit of arm showing'.

Dawkins tackles head-on such spurious arguments as that eyes could not have evolved because at earlier stages of their evolution they would not serve the purpose for which, in their fully-formed state, they exist. He reports how, as an extremely useful adaptation, eyes have evolved independently and with interesting variations about forty times in nature, in each case in very short evolutionary time-spans, and according to simple principles easily replicable by a computer programme. One of the most convincing-seeming of the anti-evolutionists' arguments thus crumbles, as they all do, at the touch of knowledge.

Dawkins' account of the way religious beliefs persist in human societies is very telling. Beliefs mimic the way viruses infect hosts, using them to replicate and spread further. Viruses, and indeed parasites of any kind, require a potential host to have two friendly characteristics. One is a readiness to replicate information accurately, and the other is an aptness to obey instructions contained in that information. Biological cells and computers are both virus-friendly environments because by their very nature they embody these two characteristics. But so do human brains, and especially the brains of the young, which in order speedily to acquire language and a massive body of information about the social and natural worlds around them, have to be open, receptive, gullible, plastic, and trusting. 'Like immune-deficient patients,' Dawkins writes in a striking passage, 'children are wide-open to mental infections that adults might brush off without effort,' making them 'easy prey to Moonies, Scientologists and nuns'.

And this is why almost all children follow the religion of their parents rather than any of the many other religions available. It is not the nature of the deity, the beauty of the liturgy or the stained-glass windows, the quality of the moral teaching, the profundity of the metaphysics, or any other property of a given faith, that attracts the vast majority of its adherents, but whether they were indoctrinated into it as intellectually vulnerable children when they were unable to defend themselves against that indoctrination.

'If you have a faith,' Dawkins writes, 'it is statistically over-whelmingly likely that it is the same faith as your parents and grandparents had . . . By far the most important variable determining your religion is the accident of birth. The convictions that you so passionately believe would have been a completely different, and largely contradictory, set of convictions, if only you had happened to be born in a different place.' And this is a mere fact of epidemiology.

The stark contrast in play is, of course, with science. Your place of birth and the outlook of your parents has no effect on what your use of scientific method will reveal about aspects of the natural world. The value of Planck's constant does not vary according to whether it is measured by a Hindu or a Muslim, as does (for example) the different amounts of flesh it is permissible for women in either religion to display publicly. Such a consideration by itself marks the difference between objectivity and subjectivity, the rational and the non-rational, the neutral and the tendentious, and so on for the family of contrasts which distinguish science and reason from the ancient grip of what passed for science in the day of the caveman.

Dawkins is necessary to our culture; like the gadfly Socrates he repeatedly seeks to sting the public mind out of the inertia which allows religion's dangerous and regressive influence to corrupt humanity's best efforts to understand the universe, and to know the truth about it and ourselves.

TO RISE ABOVE

Marek Kohn

UNDERSTANDING what Margaret Thatcher means to Richard Dawkins is the key to understanding what society means to him. Misunderstanding his relationship to her is the basis of a widespread misreading of his place in the political scheme of things. One way and another, the woman has haunted him since she and he rose to prominence in the mid-1970s.

Their ascents were not unconnected. Dawkins took the opportunity to begin what became *The Selfish Gene* during power cuts that ensued from disputes between the coal miners' union and the Conservative government, led by Edward Heath. These struggles eventually led to Heath's electoral defeat, which then led to his replacement as Conservative leader by Thatcher in 1975, the year during which Dawkins completed his book.

Ever since then, the two have been linked by observers who have perceived a deeper connection between the fortunes of the biologist and the former chemist. 'At the same time that Ronald Reagan and Margaret Thatcher preached that greed was good for society, good for the economy, and certainly good for those with anything to be greedy about, biologists published books in support of those views', writes the primatologist Frans de Waal. 'Richard Dawkins's *The Selfish Gene* taught us that since evolution helps those who help themselves, selfishness should be looked at as a driving force for change rather than a flaw that drags us down.'[1] For de Waal, then, *The Selfish Gene* was a tributary to the great current of neoliberal ideology that swept through the world in the last quarter of the twentieth century.

Dawkins has also been presumed to share Thatcher's antipathy to the welfare state. According to Steven Rose, Richard Lewontin, and Leon Kamin, 'Dawkins . . . criticizes the "unnatural" welfare state'. They quote his remarks on how 'the privilege of guaranteed support for children should not be abused', and on the possible culpability of institutions and leaders that encourage such behaviour.[2]

The main misrepresentation here does not lie in the selective quotation, which omits his indirect endorsement: 'I think that most of us believe the welfare state is highly desirable.'[3] Nor is it the implicit idea that to object to the abuse of a system is to criticize the system itself. When Rose and his colleagues made their comments, talk of responsibilities was regarded on the left as a tactic to undermine rights. The light in which Dawkins' politics are seen has since changed. Among other things, the unions can no longer put the lights out, and one's attitudes towards them no longer determine whether or not one is seen to be on the left. These days, it suffices that he has 'always voted either Labour or Liberal, or whatever Liberal happens to be called at any particular time'.[4]

It is not just a matter of assigning him to his proper place on the political spectrum. To see where Dawkins stands, it is necessary first and foremost to grasp his attitude towards nature. In implying that he criticized the welfare state because it is unnatural, the authors of *Not In Our Genes* turned his world view upside down. Richard Dawkins observes the universe and finds 'no design, no purpose, no evil and no good, nothing but blind pitiless indifference'.[5] Only intelligent beings can introduce good, or evil, into the universe. Goodness is unnatural.

As we are the only beings we know to be capable of good or evil behaviour, our moral advances are uniquely precious. Dawkins looks at the universe and sees an infinite indifference; he looks at human society and he sees 'a delicate, gossamer structure of trust and co-operation'.[6] This, he believes, is genuinely to our credit. The final sentence of the first edition of *The Selfish Gene*—'We, alone on earth, can rebel against the tyranny of the selfish replicators'—was not a feel-good sentiment tacked on with an eye

to sales. It was a concise expression of his fundamental belief that
what makes humans special is their ability to transcend 'naked
Darwinism'.

That delicate transcendent structure has been painful to create
and is 'continuously vulnerable to what I would see as degener-
ation to naked, raw Darwinism'. A recent example comes readily
to his mind as an illustration of what naked Darwinism would be
like. It would be 'a kind of Thatcherite society'.[7] For Richard
Dawkins, Margaret Thatcher represents what we are put in this
world to rise above.

His radical separation of nature and ethics places him squarely
in the tradition established by Thomas Henry Huxley, the dominant public scientist of Darwin's day. Like Huxley, Dawkins is a
Darwinian biologist whose values are explicitly anti-Darwinian
(though his Darwinism is genuine, in its commitment to the
mechanism of natural selection which Huxley could never bring
himself fully to embrace). It is possible to contemplate the ideo-
logical landscape and conclude that *The Selfish Gene* was inevitably destined to flow into the neoliberal current, but that would be
the reverse of its author's intentions.

The Huxleyan divide also puts him at odds with contemporary
evolutionists like Frans de Waal, who see human morality as a
development of evolved psychological traits, related to ones observable in our closest primate relatives. Dawkins' dramatic alternatives—human decency or naked Darwinism—are not posed for
effect. He needs the gap between them, because he wants humans
to be special. The greater the difference between how we conduct
our relations and how our relatives conduct theirs, the more credit
we can claim for human progress.

The buzz of controversy that surrounds Dawkins in public has
obscured the curious likelihood that on questions of evolved
psychology, his sympathies may be closer to those of the human-
ities graduates with whom he shares the op-ed pages than to those
of his fellow sociobiologists. He was provoked to 'stand up and be
counted' as one of the latter by *Not In Our Genes*, 'much as I have
always disliked the name'.[8] He seems to have always disliked the

implications for our species too. Loath to develop arguments based on the idea of an evolved human nature, he becomes uncharacteristically hesitant when the subject comes up. If prompted he may begin to draw speculatively upon the literature, but his heart just doesn't seem in it.

Part of the reason that such lines of thought seem unappealing to him may be that they blur the Huxleyan distinction between 'is' and 'ought'. If not from God or from nature, where can we derive our sense of how we ought to live? Dawkins is unsure. His ardent moral certainty is not all it appears. His moral sentiments depend not on any system of moral absolutes but upon their own insistent power.

Above all, they are a matter of fairness. 'If I feel that somebody has been unjustly treated—overlooked for promotion or something of that sort—I have sleepless nights.' The same arresting sensations govern his reactions to perceived unfairness whether in his own circle or on the world stage. When George W. Bush first became President of the United States, after a controversial poll and by the decision of the Supreme Court, it was the kind of 'injustice' that Dawkins feels as 'an almost visceral pain'.[9] His politics arise from his conviction that fairness should be defended in public life as in private.

Tactically, his arguments rely heavily upon exposing inconsistencies in those of his opponents. This not only suits his lawyerly intelligence, but is also a way of retaining the initiative against adversaries who enjoy the comfort of absolute moral certainties. Utilitarianism has served Dawkins well since he replaced religion with it in his youth, but deciding how to pursue the greatest good for the greatest number entails a workload avoided by those with clear dogma to guide them. For a person in Dawkins' philosophical position, intense moral sentiments are especially important as a means to reach decisions.

They may, however, be constrained by the necessity to protect those delicate cooperative structures whose value Dawkins holds so high. In 2002, he signed an open letter suggesting that European agencies might withhold academic funds as part of a response to

Israel's 'violent repression against the Palestinian people',[10] but was 'dismayed' to find that he was subsequently associated with calls to boycott Israeli scientists. With three Oxford colleagues, he formed a study group to consider whether scientific boycotts are ever justified. They described the principles of the 'universality of science' laid down by the International Council for Science, and noted that these principles had survived seventy turbulent years of world history.[11] This is just the kind of structure that Dawkins cherishes and believes is worth defending. The report concluded that scientific boycotts might be justified, but only in extraordinary circumstances.[12] The three distinctive features of Dawkins' ethical response to an issue are all present in this episode: an impassioned moral sentiment against perceived injustice, a concern to support structures of cooperation, and a conclusion that is strong but not absolute.

It is easy to recognize Dawkins as a classic English liberal. He wants to live in 'the kind of society in which people don't cheat and lie, and where everybody pays their taxes' because people believe that cheating is wrong and sharing for the common good is right, not because they are afraid of getting caught 'and have no sense that the truth is something valuable'.[13] In considering what makes a society a good society, his first concern is for reciprocity, rather than the opportunities available for individuals to pursue their own interests; and his anxiety is that a society based on a sense of the common good is always vulnerable to subversion by self-interest. It may seem more liberal in spirit to refer to tax-paying than to the bearing of children without the means to support them, but the principle is the same.

A distinctive feature in his liberalism is his concern for the welfare of animals, inculcated in his childhood by his mother—and by the *Dr Dolittle* books. Another influential experience in his earlier life was the period he spent at the University of California's Berkeley campus from 1967 to 1969, joining the protests against the Vietnam war for which Berkeley's students became famous around the world, and working for the liberal Senator Eugene McCarthy's campaign to become the Democrats' presidential candidate. Berkeley did not radicalize Dawkins; but it did move him to the left of centre.

His extracurricular campus schooling seemed to resurface several decades later, provoked by George W. Bush's drive to invade Iraq. 'Please commission an opinion poll asking the electorate the following simple question,' he requested the *Guardian* '—which would you prefer: regime change in Baghdad, or regime change in Washington?'[14] The peremptory address, the reckless equivalence, the reflexive anti-Americanism, the plain but muscular phrasing, the peevish undertone: all combine to form a reminder that Dawkins occupies the niche left vacant by J. B. S. Haldane more than forty years ago. Although Dawkins scandalizes by espousing atheism rather than the Marxism that Haldane waved like a red rag before the bull of contrary opinion, the effect is to place him in a similar position in public life.

Dawkins has political opinions, animated by moral sentiments, but he has no head for politics. His views are determined by abstract principles rather than by experience of political forces in action—or of the world, for apart from the Berkeley excursion, he has spent his adult life in Oxford—and he is no scientist-politician in the Huxley mould, or even a committee man. In the latter respect he is not a fully functional member of the Establishment, and the decidedly personal character of his views enhances the impression of semi-detachment. Politicians decide on what is to be: Dawkins has left himself free to concentrate upon what ought to be.

© Marek Kohn 2006.

ENDNOTES

1 Frans de Waal, *Our Inner Ape: The Past and Future of Human Nature* (London: Granta, 2005), 21.
2 Steven Rose, R. C. Lewontin and Leon J. Kamin, *Not In Our Genes: Biology, Ideology and Human Nature* (London: Penguin, 1990), 8.

3 Richard Dawkins, *The Selfish Gene* (Oxford: Oxford University Press, 2nd edn., 1989), 117.

4 Interview with Richard Dawkins, 8 January 2001.

5 Richard Dawkins, *River Out of Eden: A Darwinian View of Life* (London: Weidenfeld & Nicolson, 1995), 133.

6 Marek Kohn, *A Reason for Everything: Natural Selection and the English Imagination* (London: Faber, 2004), 323.

7 Kohn, *A Reason for Everything* (2004). See also the evolutionist: Richard Dawkins (http://www.lse.ac.uk/collections/evolutionist/dawkins.htm).

8 Ullica Segerstråle, *Defenders of the Truth: The Battle for Science in the Sociobiology Debate and Beyond* (Oxford: Oxford University Press, 2000), 192.

9 Kohn, *A Reason for Everything* (2004), 320.

10 Patrick Bateson et al., letter, *Guardian*, 6 April 2002.

11 Colin Blakemore et al., letter, *Guardian*, 17 December 2002.

12 Colin Blakemore et al., 'Is a scientific boycott ever justified?', *Nature*, 421 (2003): 314.

13 Blakemore, 'Is a scientific boycott ever justified?' (2003), 321.

14 Richard Dawkins, letter, *Guardian*, 18 February 2003.

WHAT THE WHALE WONDERED: EVOLUTION, EXISTENTIALISM, AND THE SEARCH FOR 'MEANING'

David P. Barash

I T is not really surprising that Douglas Adams, a great fiction writer, and Richard Dawkins, a great non-fiction writer, were good friends.

At one point in Adams' hilarious *A Hitchhiker's Guide to the Galaxy*, a sperm whale plaintively wonders to itself: 'Why am I here? What is my purpose in life?' as it plummets toward the planet Magrathea. This appealing but doomed creature had just been 'called into existence' several miles above the planet's surface because a nuclear missile, directed at our heroes' spaceship, was inexplicably transformed into a whale via an 'Infinite Improbability Generator'. As Richard Dawkins has emphasized so effectively, evolution, too, is a kind of improbability generator, although its range of outcomes is considerably more finite. Because of this, there follows another, considerably more melancholy fact: after being called into existence by natural selection, human beings have no more purpose in life than Adams' naive and ill-fated whale, whose blubber was soon to bespatter the Magrathean landscape.

First, nobody gets out of here alive. This is pure biology. And at the other end, nobody arrived here except because of a chance encounter between a particular sperm and a particular egg. Had it been a different sperm, or a different egg, the result would have been a different individual. Biology again. Finally, as to *why* we are here, the life sciences once again have an answer: human beings, like all other beings, aren't here for any reason whatsoever, certainly for no purpose that in any way transcends what their genes were up to in the first place. And this is where Dawkins has been especially persuasive, clarifying for scientist and layperson alike the simple, yet often fiercely resisted fact that evolution is a genetic process, and that *all* bodies have been 'created', like Adams' Magrathean whale, for no purpose—except the dissemination of those genes.

Admittedly, there isn't much in gene propagation itself to make the heart sing. And in an increasingly overcrowded world there is much reason to deny its prodding. Moreover, no one likes to be manipulated by someone else, even when that 'someone else' is our own DNA! At the same time, as Richard emphasized so dramatically at the end of *The Selfish Gene*, it is well within the human repertoire to rebel against our evolutionary purpose(lessness), thereby saying 'No' to our genes.

Homo sapiens is probably the only life form with this capability and, indeed, the human search for meaning has been as persistent as it is inchoate. Often it takes the form of religious faith, and here I especially want to applaud Richard for his forthright acknowledgment of the basic *incompatibility* between religion and science. The easy route is to argue otherwise, claiming 'nonoverlapping magesteria', or similar drivel. Yet the reality is that religion does not stick to a separate magesterium: it regularly makes truth claims that infringe on science. And whenever religion (belief without evidence, plus dogma) has infringed on science (conviction based on evidence, plus rational theory), the former has ultimately been forced to retreat. Some argue that a final, secure outpost for religion will be precisely the realm of meaning, the claim that whereas science can tell us what happens, only religion can tell us *why*. But this is patently false. Science indeed tells us why things happen: because of thermodynamic, electromagnetic, or gravitational forces, selection pressure, and so forth, including, in many cases, a hefty dose of chaos. It also tells us that—much as many people would like it to be otherwise—the conjunction of certain types of matter known as sperm and egg, nucleotides and proteins and carbohydrates, and a very large number of other physical entities generate us, with nothing approaching 'purpose' anywhere to be seen.

Poets, to be sure, have looked for it, and some—in my opinion, the most honest—have acknowledged its absence, perhaps none more forthrightly than Heinrich Heine in his poem, 'Questions'. Heine tells of a man who asked the waves, 'what is the meaning of Man? Whence did he come? Whither does he go? Who dwells up there on the golden stars?' And in response:

The waves murmur their eternal murmur, the wind blows, the clouds fly, the stars twinkle, indifferent and cold, and a fool waits for an answer.[1]

Where, then, does this leave the search for meaning? I see two fundamental possibilities. On the one hand, we can delude ourselves, clinging to the infantile illusion that some One, some Thing, is looking over us, somehow orchestrating the universe with each of us personally in mind. Or we can face, squarely, the reality that life is meaningless.

This does not imply giving up; quite the contrary. Here again, we can learn from Richard Dawkins as well as the late Douglas Adams. Adams (and Dawkins as well) worked hard to promote public awareness of the planetary threats to species diversity. Much to his credit, Richard has vigorously opposed the Bush administration (whose policies, incidentally, are purportedly based on biblical scripture). This is the really interesting locus of compatibility: between a recognition of life's biologically based meaninglessness and another recognition, of the responsibility for people to achieve meaning in their lives—not by hiding behind the dictates of dogma, or the promise of a 'greater purpose', but by how they choose to live their lives in a world that is altogether lacking in purpose.

Call it a kind of evolutionary existentialism. In an absurd, inherently meaningless world—our unavoidable evolutionary legacy as material creatures in a physically bounded universe—the only route to meaning is to achieve it by how we engage our own sentient existence.

The vision of life's absurdity is not surprising. Indeed, it is altogether appropriate, given that human beings—just as all other living things—are the products of a mindless evolutionary process whereby genes joust endlessly with other genes to get ahead. 'Winners' are simply those who happen to be among those left standing at the present time, but how shallow that the only 'goal' is to stay in the game as long as possible! Moreover, it is ultimately a fool's game, in which we and our DNA can never cash in our chips and go home.

Death, as the existentialists insist on pointing out, makes life absurd. The only thing more absurd is to deny the absurdity, to be stuck in a meaningless life, recognizing the meaninglessness only dimly, if at all, pretending that Mommy, or Daddy, or Jehovah, or Allah, or Brahma, has everything planned out, just for us.

In his celebrated and influential book, *Natural Theology*,[2] William Paley wrote as follows about cosmic beneficence and species centrality: 'The hinges in the wings of an earwig, and the joints of its antennae, are as highly wrought, as if the Creator had had nothing else to finish. We see no signs of diminution of care by multiplication of objects, or of distraction of thought by variety. We have no reason to fear, therefore, our being forgotten, or overlooked, or neglected.' A few decades earlier, in 1785, Thomas Jefferson, on hearing of the discovery of mammoth bones, remarked: 'Such is the economy of nature, that no instance can be produced of her having permitted any one race of animals to become extinct.' The moral? Don't lose heart, fellow human beings! Just as there are thirty different species of lice that make their homes in the feathers of a single species of Amazonian parrot, each of them doubtless put there with *Homo sapiens* in mind, we can be confident that our existence is so important that we would never be ignored or abandoned. An accomplished amateur paleontologist, Jefferson remained convinced that there must be mammoths lumbering about somewhere in the unexplored arctic regions; similarly with the giant ground sloths whose bones had been discovered in Virginia, and which caused consternation to Jefferson's contemporaries.

In his famous discourse on the different kinds of causation, Aristotle distinguished, among other things, between 'final' and 'efficient' causes, the former being the goal or purpose of something, and the latter, the immediate mechanism responsible. Evolutionary biologist Douglas Futuyma has accordingly referred to the 'sufficiency of efficient causes'. In other words, since Darwin, it is no longer useful to ask 'Why has a particular species been created?' It is not scientifically productive to assume that the huge panoply of millions of species—including every obscure soil micro-organism and each parasite in every deep-sea fish—exists with regard to and somehow because of human beings. Similarly, it is no longer useful to suppose that we, as individuals, are the center of the universe, either. Efficient causes are enough.

'We find no vestige of a beginning,' wrote pioneering geologist James Hutton, in 1788, 'no prospect of an end.' For some, the prospect is bracing; for others, bleak, if not terrifying. Pascal, gazing similarly into a vastness devoid of human meaning or purpose, wrote: *'Le silence éternel de ces espaces infinis m'effraie'* ('The unending silence of these infinite spaces frightens me').

Of course, maybe I am wrong, and Hutton too, and also Darwin, and Copernicus, and Dawkins. And maybe the whale of Magrathea was on to something. Maybe each of us is genuinely central to some cosmic design. Many people contend that they have a personal relationship with God; for all I know, maybe God reciprocates, tailoring his grace to every such individual, orchestrating each falling sparrow and granting to every human being precisely the degree of importance that so many crave. Maybe we have a role to play, and maybe—as so many people in distress like to assure themselves—they will never be given more than they are capable of bearing. Maybe we aren't Magrathean whales after all, flopping meaninglessly in a foreign atmosphere, doomed to fall. And maybe, even now, in some as yet undiscovered land, there are modern mastodons, joyously cavorting with giant sloths and their ilk, testimony to the unflagging concern of a deity or, at minimum, a natural design, that remains devoted to all creatures . . . especially, of course, ourselves. But don't count on it.

On his way down, our smiling cetacean marvels at the large flat thing approaching him very, very quickly. Among his last thoughts: 'I think I'll call it "ground". I wonder if it will be my friend.' Also, it turns out, there were actually two missiles thus transformed, each with comparable improbability. While one became our philosophically inclined whale, the other turned into a pot of petunias, which thought to itself, 'Oh no, not again!' (If the whale is Adams' paean to evolutionary absurdity, the petunias are, in a sense, a cheery nod to Buddhist and Hindu reincarnation.)

In the right hands, there can be genuine humor in life's meaninglessness. Of the now vast 'literature of the absurd'—much of it theater—a large percentage is downright funny: black humor, to be sure, but humor nonetheless—including, most notably, Samuel Beckett. As Beckett's novel, *Murphy*, draws to its antic conclusion, we are given a memorable account of what ultimately became of the hero's ashes, and thus, of human goals and aspirations more generally. (It is said that Beckett's mentor, James Joyce, was so fond of this passage that he committed it to memory.)

Some hours later Cooper took the packet of ash from his pocket, where earlier in the evening he had put it for greater security, and threw it angrily at a man who had given him great offence. It bounced, burst, off the wall on to the floor, where at once it became the object of much dribbling, passing, trapping, shooting, punching, heading and even some recognition from the gentleman's code. By closing time the body, mind and soul of Murphy were freely distributed over the floor of the saloon; and before another dayspring greened the earth had been swept away with the sand, the beer, the butts, the glass, the matches, the spits, the vomit.

Beckett's best-known work, *Waiting for Godot*, is constructed of similarly shocking and occasionally brilliant absurdities. It, like *Murphy*—and like evolutionary biologists with a 'Dawkinsian' perspective—confronts the reality of meaninglessness. Two tramps, Vladimir and Estragon, spend all of Act I waiting for Godot, with whom they believe they have an appointment. He doesn't show up. Nor does he appear in Act II, leading to one wag's description of the play as one in which nothing happens. Twice.

Toward the end, Vladimir cries to his buddy, Estragon: 'We have kept our appointment, and that's an end to that. We are not saints, but we have kept our appointment. How many people can boast as much?' Vladimir replies: 'Billions'.

After all, there is nothing unusual in keeping one's appointment. We all do it, insofar as we exist. Our friend the whale similarly showed up for his appointment with the ground. (Woody Allen once noted that 99 per cent of life is just showing up; biologically, it is 100 per cent.) Maybe the key, then, is what you do while you are waiting. This is precisely what underlies the literature of existentialism: the prospect of redemption, of achieving meaning via meaningful behavior, even though—or rather, especially *because*—in the long run any action is meaningless. One of the greatest such accounts, and one of the best examples in any novel of people achieving meaning through their deeds is Albert Camus' *The Plague*, which describes events in the Algerian city of Oran during a typhoid epidemic. *The Plague* is a 'chronicle' compiled by the heroic Dr Rieux, in order to 'bear witness in favor of those plague-stricken people; so that some memorial of the injustice and outrage done them might endure; and to state quite simply what we learn in a time of pestilence: that there are more things to admire in men than to despise'.

But Rieux, and Camus himself, are quite aware that we live always in a time of pestilence. In Camus' most famous essay, 'The Myth of Sisyphus', Sisyphus stands for all humanity, ceaselessly pushing our rock up a steep hill, only to have it roll back down again. Over and over, generation after generation, death after life after death after life, that is our lot—and that of our genes. Camus concludes his essay with the stunning announcement that 'One must imagine Sisyphus happy', because he accepts this, defining himself—achieving meaning—within its constraints. Upon reflection, it is clear that Camus' stance, in which meaning is not conveyed by life itself but must be imposed upon it, is not only consistent with an informed biological perspective, it actually makes sense only *because* of that perspective.

The quest for meaning itself only takes on meaning—indeed, it only takes place at all—because people are creatures whose lives, embedded in biology, are fundamentally lacking in meaning. The greatest triumphs of existence arise from human beings struggling to make sense of what is, biologically, a purposeless world: not from attempting to rise above their biology (for that cannot be done), but from riding their biology to greater heights. Many, naively hopeful like the whale of Magrathea, simply plummet to their deaths.

At the conclusion of *The Plague*, while the citizens of Oran are celebrating their 'deliverance', Dr Rieux knows better. He understands that in the game of life, all victories are temporary, which renders his perseverance all the more grand: 'He knew that the tale he had to tell could not be one of a final victory. It could be only the record or what had had to be done, and what assuredly would have to be done again in the never ending fight against terror and its relentless onslaughts, despite their personal afflictions, by all who, while unable to be saints but refusing to bow down to pestilence, strive their utmost to be healers.'

For many, connecting evolutionary biology and existentialism seems oxymoronic in the extreme. I maintain, by contrast, that they are natural allies. (Like Richard Dawkins and Douglas Adams.) In italicizing our lack of inherent meaning while striving to respond to our very human, biologically-generated need for it, practitioners are, in their own way, refusing to bow down to some of evolution's more unpleasant imperatives. Although no one can announce final victory, Sisyphus—along with Charles Darwin, and Douglas Adams, and Richard Dawkins—might well applaud such efforts.

ENDNOTES

1   I thank Latha Menon for bringing this quotation to my attention.
2   William Paley, *Natural Theology* (1803).

WRITING

RICHARD DAWKINS AND
THE GOLDEN PEN

Matt Ridley

BEFORE *The Selfish Gene*, scientists wrote books for each other, or for laymen, but rarely for both. The great interpreters of science, such as Peter Medawar, J. B. S. Haldane, or Arthur Eddington might write with fluency, wit, and verve, but they were still more inclined to explain established ideas than to explore new mysteries. Gracefully and graciously they gave you the answer rather than the argument. More than anybody before him, Richard Dawkins thought that if he was to persuade his fellow scientists of a new truth that seemed to him 'stranger than fiction' he might as well try to enlighten the rest of us while he was at it. The result was that he gave laymen a chance to eavesdrop on scientific debate in action. He was quite explicit about it in his preface: 'Three imaginary readers looked over my shoulder while I was writing, and I now dedicate this book to them. First, the general reader, the layman . . .' The other imaginary readers were the expert and the student.

I think that is why I still recall a sense of slight bewilderment when I read the newly published book as a first-term undergraduate at Oxford. Was this chap's theory right or not? Until now my teachers had helpfully divided the world of science into right and wrong ideas. But here, I suddenly realized, I was going to have to make up my own mind. The handrails had gone.

It would be wrong to claim that modern popular science writing sprang fully formed into the world in 1976. Non-fiction publishers had been given plenty of warning that science was a rich vein to mine, evolutionary theory in particular. In 1961 the screen writer Robert Ardrey's *African Genesis* popularized the killer ape

theory. Konrad Lorenz followed with *On Aggression* in 1963 and
Desmond Morris with *The Naked Ape* in 1967. The latter would
sell more than ten million copies. The great success of Jim
Watson's *The Double Helix* (1968), Jacques Monod's *Chance and
Necessity* (1970) and Jacob Bronowski's *The Ascent of Man*
(1973) reinforced the view that science could generate a huge
best-seller almost every year. In this context, *The Selfish Gene* was
merely the 1976 incarnation of a regular phenomenon.

But it stood out in two ways. One was the sheer brilliance of the
prose. Dawkins' sentences had such rhythm, his words had such
precision and his thoughts had such order that his book was tasty
literature as well as nourishing argument. Undoubtedly for this
reason, by word of mouth the book gathered pace among lovers of
writing as well as connoisseurs of science—and among women as
well as among more fact-obsessed men. The other exceptional
thing about *The Selfish Gene* was its argument, which was to many
people brand new, utterly unexpected, deeply unsettling, and yet
plainly unsettled. In other words, not only could readers feel privy
to an unfinished debate, but they could see the world in a different
way. Where Ardrey, Lorenz, and Morris had told them some surprising things about themselves, Dawkins turned their whole world
upside down. If he was right, he had found a great new truth that
penetrated to the heart of human and biological existence. For
example, most people take it for granted that parents are generous
with time, work, and money to their children. They do not even
stop to wonder why. Now came an extraordinary explanation of
why—instead of how—they were so altruistic: genes that caused
adults to invest in their offspring had spread within the species at
the expense of genes that caused indifference. In a sense, therefore, our most benevolent acts could be recast as selfish ones at
the level of the gene thus, paradoxically, explaining why we were
not always selfish. 'We are survival machines—robot vehicles
blindly programmed to preserve the selfish molecules known as
genes.'

Throughout the rest of his career, the argument would continue
to overshadow the writing skill. Dawkins is usually reviewed and

discussed in terms of what he has to say rather than how he says it. He is rarely judged purely as a writer. This is a tribute to his passion for truth. Yet he is a craftsman of exceptional skill and would only get better in subsequent books, especially *The Blind Watchmaker*, the book in which he drives a juggernaut of logic through the argument from design. The combination of exceptional writing talent and controversial argument would influence a generation of writers who came after. No longer was it sufficient to deliver the facts *ex cathedra* to a grateful public. Nor was it necessary to stick to unemotional prose. Science writers now reminded themselves they were writers and their currency was words; poetic flights of fancy, ample use of metaphor, and personal appeals to the reader all became more common.

Against group selection, intelligent design, punctuated equilibrium, and God, there was always an argument to be won and Dawkins knew you do not win arguments by fact alone. You must sell your case, dress it up in ways that appeal to the feelings as well as the thinkings of your reader, and keep the reader turning the pages. The old advertising adage has it that you sell the sizzle, not the sausage. To do this without sacrificing the truth of your case is not easy, and few can manage it, but Dawkins has always been able to make his truths sizzle—without getting his facts wrong.

Few scientists appreciate this point. They tend to believe that truth prevails in arguments—despite ample evidence that various opponents of science, from anti-GM activists to creationists, are able to market their wares effectively by selling sizzle, however poor their sausage. Likewise, the triumph of gene-centred evolutionary thinking was by no means inevitable, however brilliant the algebraic proofs supplied by Bill Hamilton and the ingenious arguments of Robert Trivers. Dawkins' prose did an enormous amount to convert people to this fresh way of thinking. Yet good writing can make itself invisible. When somebody writes as smoothly as Dawkins, he makes it look easy, and colleagues tend to think he has 'just' written a popular book. It is no secret that Dawkins, despite the complexity of the arguments in *The Extended Phenotype*, stood accused of being a scientific lightweight among his professional colleagues for having written in an accessible style.

An unexpected effect of the success of *The Selfish Gene* was to revive the central role of the book as a scientific art form. The 1970s saw a surge of interest in science coverage by newspapers and magazines as it dawned on journalists that a revolution in computing and in biotechnology was breaking and needed explaining. It also saw a surge in television coverage of science. It seemed at the time that television would gradually conquer the field, but who could have predicted that Bronowski's series on 'The Ascent of Man' would prove to be the high-water mark of television science and that by 2005—natural history aside—the science blockbuster series on television would be all but extinct. Still less, who could have foreseen that, to survive in a desert of reality TV, regular science programmes would be reduced to ever more frantic attempts to inject distracting visual thrills into content-light stories about forensic science, natural disasters, and child development? The truth is that most scientific arguments are ill-suited to television, since they require thought, detail, and argument, three things the medium detests. Their natural home would continue to be the book.

However, it has to be said that you search in vain for an explosion of great popular science books immediately after 1976. Stephen Jay Gould's first book of essays *Ever Since Darwin* was the phenomenon of 1977, but though Dawkins and Gould continued to sell mouth-watering quantities of successive books in the 1980s it was not until 1987 that popular science again saw a best-seller on the scale of *The Selfish Gene*. That year saw James Gleick's *Chaos* rocket into the best-seller lists, followed the next year by the even more extraordinary success of Stephen Hawking's *A Brief History of Time*. Gleick and Hawking, like Dawkins, were telling the public about new scientific ideas in ways that engaged them as equals. But they were not simultaneously trying to argue with rival professionals. They were in the explaining, not the exploring, tradition. Not until Steven Pinker's *The Language Instinct* in 1994 did a true successor to

The Selfish Gene appear: an argumentative book aimed at persuading professional scientists as well as enlightening laymen and written in unputdownable prose.

Publishers are herd animals. The success of Hawking in particular sparked a search for the next science best-seller that saw an unprecedented wave of popular science titles being published throughout the 1990s (not to mention an obsession with entitling books 'brief history' or 'short history'—Bill Bryson would hit the jackpot again with the phrase in 2003). Many of us were grateful beneficiaries of the resulting largesse. Suddenly science writers could command large advances, egged on by an aggressive new breed of literary agent led by John Brockman. Dawkins himself promptly produced three wonderful books in four years—*River Out of Eden*, *Climbing Mount Improbable*, and *Unweaving the Rainbow*. Even as television gave up on many aspects of science, and newspapers retreated to covering technology, environmental scares, and popular psychology, the book trade was making stars of scientists. People like Richard Leakey, Jane Goodall, Carl Sagan, and Steven Pinker could fill the biggest lecture halls as Thomas Henry Huxley had once done. But booms usually end in busts and the boom in popular science publishing was no exception. The deflation began after the debacle of Murray Gell-Mann's *The Quark and the Jaguar* (1994), which Brockman sold for an advance of $550,000 for American rights alone. Gell-Mann returned much of the advance after failing to complete the book as promised, though it did eventually come out. The next year the phenomenon was *Longitude* by Dava Sobel and narrative non-fiction, rather than argument, became all the rage. Cod, tulips, salt, and zero were the themes of the moment, not to mention, on a grander and more analytical scale, guns, germs, and steel.

In recent years, even as his imitators swarmed, the master continued to dominate the lists. *A Devil's Chaplain* and *The Ancestor's Tale* effortlessly climbed the charts. Others may aspire to his facility with words, or reach for his ease with literary allusion and metaphor, but they can only dream of his ability to change the way the scientific world thinks by means of a popular best-seller.

EVERY INDICATION OF INADVERTENT SOLICITUDE

Philip Pullman

I THINK it's fair to guess that most of Richard Dawkins' many readers are not using *The Selfish Gene* and its successors as textbooks to help them pass science exams. That he is a highly distinguished scientist is not in question, but many scientists have achieved great distinction—and indeed written textbooks—without once writing a popular best-seller.

Nor is it likely that a large part of his readership has mistaken their nature, and gone to his books in search of evidence that Charles Darwin, contrary to popular belief, was a secret member of the Templars and spent his life tracing the blood-line of Jesus Christ by means of the secret code embedded in the shells of Galapagos tortoises.

So what is it that so many readers are responding to in his work?

We might start by looking at the best-seller lists, and seeing what it is that people like in the books they buy in large quantities. However, the secret of best-sellerdom is beyond discovery. If it were, publishing would be an easy business to make a fortune in. But best-sellers keep surprising everyone in the book trade. No one knows what the secret is: hence the story—perhaps it's apocryphal—of the finance director in a publishing conglomerate instructing the editors to publish only best-sellers in future, because last year they published several books that weren't.

So there's no formula. To be sure, there *are* books written formulaically, and some of them sell in large numbers. But among the books that have been read by millions of people, and reread, and have stayed in print, are works of genuine originality and literary power. The best-seller lists are no guarantee either of quality or of the lack of it.

So, once again: what is it about Dawkins' writing that appeals to so many readers?

I think there are a number of things that could account for it. The science might well be one: this is a great age of popular writing about science, and the appetite grows by feeding. But here I'm going to take the science as read, and look at three other possibilities.

The first is his uncommon gift for phrase-making. The ability to find happy phrases seems to be inborn; some people move confidently through the ocean of language, as free as a fish, while others have to stay near the shore and keep a nervous foot on the sea bed. It's not dissimilar to the gift for metaphor, but it involves a sensitivity to the textures and flavours of language, too, which metaphor doesn't, necessarily: such a picture as Salvador Dali's fruit-dish-that-looks-like-a-face-that-looks-like-a-beach is a metaphor that doesn't involve language until you try to describe it. *The Selfish Gene* is a metaphor, but it's not the metaphorical content alone that makes it memorable; it's the iambic rhythm, it's the intriguing combination of common word with scientific term, of word we can picture with word we can't, of word-that-implies-purpose with word-that's-bare-of-purpose, it's the echo of Oscar Wilde's *The Selfish Giant* and the subliminal hint of a story to come (who knows? Perhaps a story in which the gene will be punished for his selfishness, and repent), it's the three short vowels followed by a long one, it's all those things.

But it isn't only in his titles that we can see Dawkins' gift for combining words in a knot that stays tied. The words I've used at the head of this piece come from page 459 of *The Ancestor's Tale*, where he's writing about how plants fix atmospheric nitrogen—or rather, the plants themselves don't: 'It is symbiotic bacteria—specifically *Rhizobium*—housed for the purpose in special nodules provided for them, with every indication of inadvertent solicitude, on the roots of the plants.'

That's the sort of thing that makes me clap my hands. It's

funny. Dickens would have enjoyed writing a phrase like that. There's no need for it; it doesn't make the sense any clearer; it doesn't seem to have been struggled for, in a desperate attempt to wrestle meaning out of the grasp of incoherence; it just sprang into being, and it's delightful. Actually, the impression of effortlessness is exactly that: an impression. You can be naturally very good at that sort of thing, but you still have to work hard. The delight for the reader is in the illusion that the phrase simply flowered under the author's fingertips as he sat at the computer screen. It might have done; but there was a great deal of hard gardening that led to that flower, and I guess the author saw it bloom with pleasure. As Sherlock Holmes remarked, 'Our highest assurance of the goodness of Providence seems to me to rest in the flowers'—but perhaps we'd better not go any further with that.

The second thing I think readers respond to so strongly in the work of Richard Dawkins is that very old-fashioned thing, the revelation of a personality. The great essayists and writers of non-fiction observe their subjects with a closeness of attention that's halfway between obsession and rapture, but there's always something else that comes through the words, and that's a clear and vivid impression of the human being behind them. It doesn't have to be a charming personality, but it has to be a strong one. Furthermore, it doesn't have to be 'real'. The voice that speaks to us from the pages of an essay is as much an invention of the writer as the overtly fictional characters in a novel. (Though even that has to be qualified: the writer might not actually be aware that he's making 'himself' up.) But real or not, the character matters. The rough, curious, bustling energy and boundless eccentric knowledge of Robert Burton, the delicate courage of Charles Lamb building his ramparts of cobweb against the encircling madness, the shafts of kind-heartedness that break through the brisk rational tireless mocking nonsense of Bernard Shaw—those are the reasons for which I read them. An hour in the company of a clear and decided personality is a tonic, whether you agree with them or not.

As I say, that's an old-fashioned view, but I think a lot of readers are old-fashioned in that way. And I wouldn't be surprised if they enjoyed the character that reveals itself in Dawkins' writing for much the same reasons as I do. The scourge of the church, the flayer of the creationists, the pitiless mocker of the superstitious— plenty of people could have assumed those roles, and been soon forgotten; but because Dawkins takes the subject so seriously, he goes further, and harder, and more thoroughly than anyone who was just playing at it. Take the chapter called 'Do good by stealth' from *River Out of Eden*. He begins by quoting a letter from an American minister who had been an atheist, but who was converted by reading an article about orchids that mimicked wasps and tricked the insect into copulating with them, and thus spreading their pollen. The minister had made the mistake of assuming that such an arrangement had to be set up perfectly in advance, all at once, or it wouldn't work at all, and thus had to be the work of a divine designer.

Dawkins says, 'Others, no doubt, come to religion by different routes, but certainly many people have had an experience similar to the one that changed the life of this minister (whose identity I shall withhold out of good manners). They have seen, or read about, some marvel of nature. This has, in a general way, filled them with awe and wonderment, spilling over into reverence.' From there he goes on to disentangle the mistaken assumption with great clarity and effect, showing with the help of numerous examples that there could have been any number of intermediate steps between *no resemblance at all* and *close resemblance*, each of which would have given the plant a small but valuable evolutionary advantage. (In the course of this we meet another of his admirable phrases, when he calls the minister's letter an example of 'the argument from personal incredulity'.) He concludes the chapter by taking in 'the creationist's favourite conundrum. What is the use of half an eye?' and demolishing that too, with a zest that reminds me of a great batsman in his prime—Viv Richards, say—dismissing the bowling majestically to all parts of the ground and not once looking in trouble.

But other batsmen have played majestically; other writers have left their opponents' arguments in ruins. Only Dawkins would add that '(whose identity I shall withhold out of good manners)'. That's the authentic touch; the asperity is shocking; we gasp, but the gasp is half in admiration at the sea-green incorruptibility of the voice, at the realization that this debate is *serious*. He *means* it.

There are those who say he goes too far in his attack on religion, that he fails to understand the true nature of religious mystery, that his criticism relies on a parody of faith that he himself has set in place and that few believers take seriously any more. Those who make that objection tend to belong to the mild, kindly end of the religious spectrum. These days, we should be in no doubt about what the other end of the spectrum is like; and we might remember that no social structure ever gives up power because it wants to. If some parts of the Christian church are decent and tolerant today, it is because the crusaders and inquisitors and witch-burners have been shamed and stripped of their authority by the great critics of religion—some, indeed, who belonged to the church itself, but all of whom were accused, in their time, of going too far.

And the vividness and integrity of character needed to maintain a stance like that are uncommon. They always were, but in these celebrity-worshipping days, they seem harder to find. The qualities that make for instant celebrity are quite different. When we come across a body of work that offers us the real thing, we know the work will last.

The third reason I think readers enjoy Dawkins is that he tells stories. They are interested in his subject matter, to be sure, but they are seduced by the way he tells them about it. Open a page of any of his books and it's very hard to stop reading. Storytelling is partly a matter of rhythm, and I don't just mean the rhythm of how the words sound in succession: I mean the pace at which new facts are introduced, new examples offered, new topics turned to. But saying what it is isn't the same as saying how it works: it isn't regular, that's the problem for analysis and generalization. It's almost too subtle to be made explicit, just as the minute and infinitely flexible displacement of rhythm called *swing* passes through the relatively coarse mesh of musical notation. We can feel it; we can recognize it at once; but to set down how it's done . . . 'If you gotta ask, you'll never know', as Louis Armstrong apocryphally said when asked what jazz was.

However, large matters of organization are a little easier to describe. One essential thing to bear in mind, when telling a long story, is the readers' need not to feel lost. Readers are happy to be mystified, as long as they think a solution is forthcoming; they are delighted to be tantalized and kept in suspense; but if they feel merely bewildered, if they just can't see why they're here or where *here* is or what's going on, the story isn't working. For example, *The Lord of the Rings* works so well as a narrative because at every point the reader knows why the characters are here, and what they're doing: they are all working to help Frodo take the ring to Mordor and drop it into the volcano. The characters may be lost, but the reader should never be; the quest may be difficult to do, but it has to be easy to understand.

So a sense of large-scale structure is an essential gift for the storyteller. The shape of Dawkins' latest book, *The Ancestor's Tale*, is a good example of his way with this. He sets out to tell the story of evolution in an unusual way: by going backwards through time towards the origin of life, rather than by coming forwards in the conventional way. Now there are good reasons for telling stories in the conventional way. If an unusual voice, or point of view, or tense, or structural device is being used, it had better express something important, or it will merely seem like an irritating affectation. In this case, the form allows Dawkins to illustrate the unity of life on earth quite brilliantly by picturing modern humans returning towards our origins in a sort of pilgrimage, and meeting our relatives the Cro-Magnons and Neanderthals, and a little later being joined by our cousins the apes, and then the monkeys, and so on.

Each meeting-place in the journey is the point at which, if the flow of time were going forward, we would find one evolutionary line diverging from another. Seeing it this way round helps us to understand how closely we are all related—we have ancestors in common, no matter how different we seem—but the 'pilgrimage' structure allows Dawkins to do something else, as well. Just as Chaucer's pilgrims told stories on their way to Canterbury, so these pilgrims tell theirs: or rather Dawkins does, with each tale illuminating some aspect of the extraordinarily rich and varied forms that life has taken, and of its meaning for us. So 'The Grasshopper's Tale' touches on the complexities and difficulties of discussing race; 'The Duckbill's Tale' takes in Nimrod early-warning aircraft and radar, the way some fish generate electric fields, and the possibility that the study of how platypus venom works on pain receptors might lead to the discovery of a new form of pain relief for cancer victims; 'The Redwood's Tale' touches on carbon dating, and the various other forms of telling how old things are, and the great value of being able to cross-check one method with another, and something called the Parsons code, which allows you to find the name of a tune you've got in your head without needing to read music—and so on.

The ingenuity is great, but here's the point: it is truly *expressive*. What it expresses is a passion for the variety and splendour of life. All these extraordinary descriptions, with every one of their myriad connecting links, are there to perform a task at which Dawkins says supernatural beliefs miserably fail: namely, 'to do justice to the sublime grandeur of the real world'.

And that's exactly what Dawkins' writing does. It's what he's been doing ever since *The Selfish Gene*, and I think that that is what readers are responding to most deeply. He is a coiner of memorable phrases; he is a ferocious and implacable opponent of those who water the dark roots of superstition. But mainly he celebrates. He is a storyteller whose tale is true, and it's a tale of the inexhaustible wonder of the physical world, and of ourselves and of our origins.

© Philip Pullman 2006.

INDEX